World Licensing and Operating Directory

A guide to getting on the air from more than 200 countries and territories around the globe

Compiled and edited by
Steve Telenius-Lowe, 9M6DXX

Radio Society of Great Britain

Published by the Radio Society of Great Britain, 3 Abbey Court, Fraser Road, Priory Business Park, Bedford MK44 3WH, England.

First published 2008

ISBN: 9781-9050-8646-7

Publisher's note
The opinions expressed in this book are those of the author and not necessarily those of the RSGB. While the information presented is believed to be correct, the author, the publisher and their agents cannot accept responsibility for consequences arising from any inaccuracies or omissions.

Cover design: Kim Meyern
Layout and design: Steve Telenius-Lowe, 9M6DXX
Proofreading: Dr George Brown, MW5ACN
Production: Mark Allgar, M1MPA

Printed in Great Britain by Page Bros (Norwich) Ltd

Contents

Introduction

Welcome to the *World Licensing and Operating Directory* and thank you for buying this book. The purpose of this book is really very simple: to encourage more on-air amateur radio operation, and from as many different places in the world as possible.

It does this in two ways. Firstly, by providing a source of information on licensing in countries around the world and thus making it easier for amateurs to operate when abroad, whether for business or pleasure. Secondly, by providing a source of information on the numerous amateur stations around the world that are available for rent by visiting licensed amateurs.

With the development of tiny, light-weight transceivers, switch-mode power supplies and even relatively light-weight high-power linear amplifiers, it has never been easier to operate from abroad. This is covered in Chapter 2 of this book.

LICENSING INFORMATION SECTION

Licensing, however, is sometimes the biggest challenge to overcome, particularly if the chosen destination is a country that does not issue licences as readily as, say, the CEPT countries or the USA. It is to be hoped that the 'Licensing Information' section of this book will show that, in some cases at least, it is perhaps not as difficult to obtain a licence as might have been assumed.

The editor and publishers are indebted to Veikko ('Veke') Komppa, OH2MCN, for agreeing to make the information on his 'Worldwide Information on Licensing for Radio Amateurs' website freely available to the editor of this book. The information on Veke's pages has been collected over a period of more than 12 years and includes contributions from some 320 individuals, all of whom are thanked.

Veke's website, excellent though it is, was only one of several sources of information for this section of the book. Many contributors have kindly provided up to date information and these people are acknowledged, with thanks, below. Where the licensing authority has its own website this was carefully checked for current information on amateur radio licensing.

RENTAL STATIONS SECTION

Getting a licence is, of course, only half of the story. Where are you going to operate from? While it can be an interesting challenge to mount a DXpedition to a remote location, taking all the necessary equipment yourself, many people prefer to opt for an easier life and instead rent an existing station.

It is anticipated that the 'Rental Stations' section of this book will appeal to two groups of people: firstly, those keen contesters and DXers looking for a competitive station in an interesting location that they can rent without having to carry transceivers, linear amplifiers, antennas and masts half way around the world. Such people will find a number of suitable stations in the pages of this book. The second group of people are those who wish to take a holiday in a particular part of the world, perhaps with their family, and who wish to complement the holiday with some amateur radio operating from an unusual or unfamiliar location. They too will find plenty of scope for their holiday planning in this book.

The 'Rental Stations' fall into several categories, including apartments with a station, hotels, private 'bed and breakfast' businesses, usually in the owner's own home, existing club stations, holiday cottage rentals etc. We have *not* included so-called 'ham-friendly hotels' (hotels where amateurs have been allowed to set up their stations by the hotel management) simply because the scope is just too enormous. After all, virtually every hotel in the world is, potentially at least, 'ham-friendly' - if the management is doing its job properly.

Instead, every place listed in the 'Rental Stations' section of this book offers the visiting amateur the use of either a rig or an antenna as an absolute minimum. The majority offer a complete station with transceiver, power supply and antennas, while a few are amongst the most sophisticated amateur stations in the world with multiple transceivers, linear amplifiers, stacked monoband beams, and a high degree of station automation.

For every station listed, the property or station owner has given his or her permission to be included in the book and has confirmed that the details given at the time of compilation (between late 2007 and mid 2008) is accurate. However, stations are upgraded, equipment is sold and replaced, antennas are blown down in winter storms or tropical typhoons, so no guarantee can be or is given by the editor or publishers of this book that the property, equipment or antennas are precisely as described. If you are attracted to a location specifically because of a particular feature described in this book, you are advised to check with the owner that it is still in place or still operational.

A word on payment for rental of these stations. Where possible we have given a guideline rental price, so you have a fair idea of the amount of money required to rent the station. However, in many cases, there are higher prices during peak seasons or major contest weekends, or alternatively discounts may sometimes be available. Again, check with the owners when booking.

No mention is made of how the property owners wish payment to be made. Some may require payment to be made by cheque (US: check) in their own currency, or by *PayPal* perhaps. Some may insist on full payment before the use of the station; others may ask for a 50% deposit being made to secure the booking. It is entirely the responsibility of the individual to negotiate these arrangements with the owners, and the editor and publishers will take no responsibility for bookings not honoured.

CURRENCIES

A note on the use of currencies throughout the book. We always use the currency quoted by the licensing authority in the Licensing Information section, or the property owner in the Rental Stations section of this book. If this is pounds sterling (£, GBP) or US dollars ($, USD), no conversion is given because the conversion rate was very close to $2.00 = £1 at the time of writing. Where other currencies are quoted, an approximate equivalent rate in pounds sterling is given in brackets, for example "SEK 350 kronor (approx GBP £27)". Simply double the pounds sterling amount for an approximate rate in US dollars.

UPDATES

While every care has been taken in compiling this book, errors will inevitably occur in a work of this scope. When researching the Licensing Information section conflicting details often came to light and an educated decision had to be taken as to which was most likely to be the more accurate or up to date. Inevitably I will have taken some wrong decisions. For a few countries (actually a remarkably small number) I was unable to unearth any details about amateur radio licensing at all.

In the Rental Stations section of the book, there are undoubtedly more stations of which I am still unaware, despite research over a period of some 10 months.

If you have current personal knowledge of licensing in any country which conflicts markedly with the information in this book, or if you are aware of any rental stations that are not mentioned (and have not been excluded for the reasons described at the beginning of Chapter 5), you are invited to contact the editor at teleniuslowe@gmail.com with details. (Veke Komppa, OH2MCN, would also appreciate updates to his 'Worldwide Information on Licensing for Radio Amateurs' website: see http://www.qsl.net/oh2mcn/license.htm for details.) Any amendments or updates to this book can be found the supporting web page www.rsgb.org/books/extra/wlod.htm

ACKNOWLEDGEMENTS

First and foremost, to Veikko ('Veke') Komppa, OH2MCN, for his cooperation in compiling the Licensing Information section of this book, sincere thanks. Without Veke's willing cooperation, the research involved would have been infinitely more tedious and the book may never have been produced.

Mark Allgar, M1MPA, at RSGB Headquarters, provided encouragement and background information that was essential to the production of this book. In the introductory chapters I have re-edited and brought up to date material I originally wrote for a now out of print edition of the *RSGB Operating Manual* and for *DXpeditioning - Behind the Scenes*. I am grateful to the RSGB and the Five Star DXers Association for permission to use this material.

I would like to thank Dr George Brown, MW5ACN, for proofreading and making corections where necessary. Any errors remaining are undoubtedly mine, not his.

I am also extremely grateful to the many other individuals who helped provide information for both the Licensing and Rental Stations sections of this book. With apologies to anyone who has been accidentally omitted, thanks go, in particular, to:

Seewoosankar 'Jacky' Mandary, 3B8CF; Natig Kasimov, 4J5T; Ranko Boca, 4O3A; Åke Rosvall, 5R8FU; Jan-François Lorne, 6W7RV; Josh Walker, 6Y5WJ; Rolando Milin, 9A3MR; Alfons Undan, 9M6MU; John Plenderleith, 9M6XRO; Allan Ming Addie, 9M8MA; Yeshey Dorji, A51AA; Dave Becker, AL7DB; Mario Iberkleid, CP1FF; Don Stewart, CT1IUW / G3TIR; Richard Serván, CX2AQ; Jorge Diez, CX6VM; Reiner Freimuth, DG7FX; Ulli Krieg, DL2AH; Carsten-Thomas Dauer, DL2OBO; Jürgen Eisinga, DL2YAG; Erwin Scherr, DL4NCF; Rev Dr Bill Burton, DU3/G4CWA; Des Clarke, E51DD; Duncan Lindsay, EA5ON; Bill Kiwitt, EA8AZC; Vincent Colombo, F4BKV; Georges Santtalikan, FG5BG; Laurent Bellay, FM5BH; Stan Wisniewski, FO5IW; Andy Chadwick, G3AB; Jeff Blight, G4SOF; Richard Everitt, G4ZFE; Darren Collins, G0TSM; Alex Gartshore, GD3UMW; Brian Sparks, GM4JYB; Phil Cooper, GU0SUP; Chris Green, GW4VAG; André Breguet, HB9HLM / CN2DX; Maj Narissara Shaowanasai, HS1CHB; Mathias Bjerrang, JW5NM; Noah Gottfried, K2NG; Charles 'Frosty' Frost, K5LBU; Ken Widelitz, K6LA; Stan Barczak, K8MJZ; Brad Zuehlke, K9BZ/KP2; Howard Olsen, KD4ML; Terry Clayton, KH6SQ; Gaynell Larsen, KK4WWW; Steve Wheatley, KU9C; Bill Walker, KX4WW; John Core, KX7YT / S21YL; Graham Dawes, M0AEP / VP2MDD; Max Cotton, M0CHQ; Tim Beaumont, M0URX; Bob Barden, MD0CCE; Phil Taylor, MJ0JER; Steve Redmond, MW0ZZK; Trey Garlough, N5KO; Finn Jensen, OZ1HET; Emily Thiel, P43E; Michael Dirksen, PA5M; Ramon Kaersenhout, PZ5RA; Per Green, SM0DFP; Lars Aronsson, SM3CVM; Jörgen Norrmén, SM3FJF; Ulla Norrmén, SM3LIV; Odd Westby, SM4SXQ; Manos Nerantzulis, SV9ANJ; Carlos Diez, TI5KD; Bob Fox, V31MD; Chris Chapman, VK3QB; Rick Rodgers, VK4HF; Mirek Rozbicki, VK6DXI; Mal Johnson, VK6LC; Jody Millspaugh, VP5JM; Ed Kelly, VP9GE; Geoff Howard, W0CG; Dr Glenn Johnson, W0GJ; Ken Eigsti, W0LSD; Jim Millner, WB2REM; Sam Harner, WT3Q; Hiroo Yonezuka, XU7AAA; Octavio Miranda, YN2N; Igor Derivolov, Z32ID; Graeme Hunt, ZL1ANH; Brian Miller, ZL1AZE; Ralph Sutton, ZL2AOH; Tomàs Anibal Zapattini, ZP5AZL; Daniel Hubbard, ZS6JR, and Mrs Esther Grant (Malawi), Ms Hiroko Tani (Api Corporation, Japan), Marty Kaiser (Cayman Islands), Luis of VIP Guest Hotel (Koror, Palau), Matt Smith (Bahamas), Miss Alison Stewart (UK), Lynn Van Leeuwen (Hawaii), and David Wilson (Philippines).

DISCLAIMER

Whilst every effort was made to ensure that the information given herein is accurate, no responsibility is accepted by the editor or the RSGB for any errors, omissions or misleading statements in that information by negligence or otherwise, and no responsibility is accepted in regard to any subsequent action based on this note.

Steve Telenius-Lowe, 9M6DXX
Sabah, East Malaysia, August 2008

So you want to go on a DXpedition?

2

The main purpose of this book is to encourage radio amateurs to operate from as many places around the world as possible and, by providing information on licensing and the availability of rental stations, to make it easier to do so. Such operation from abroad is often called a 'DXpedition' and this can probably best be defined as a portable operation from another country, DXCC entity, IOTA island or locator square, purely or mainly for the purpose of making amateur radio contacts. This chapter is mainly about operating from abroad on HF, although most of the general comments about licensing, customs, equipment and antennas apply equally to VHF, the main difference being that - other than for specialised 6m or moonbounce operations - most VHF operations are likely to be closer to home than their HF counterparts.

The phenomenon of 'DXpeditioning by the masses' is a relatively recent one. Of course, radio amateurs have always travelled, and there was a number of 'DXpedition pioneers' from the 1930s or 40s onwards who are worthy of individual mention.

However, it wasn't really until the development of relatively small 12V-operated 100W HF transceivers such as the Kenwood TS-430S or the Icom IC-735 in the 1980s that the concept of the 'one-man DXpedition' really took off. Before then, equipment was simply too big and heavy to be easily transported to remote locations.

In the quarter of a century since then, portable transceivers have continued to get smaller, lighter and more sophisticated. Kenwood led the way with the release of the TS-50S in the early 1990s, followed swiftly by Icom with the enormously successful IC-706 and its later variants. Now, as the decade of the 2000s is drawing to a close, Yaesu currently offers the smallest and lightest 100W HF transceiver on the market, the FT-857D, which weighs in at just 2.1kg. Arguably the most sophisticated 'mini-mobile' available is Icom's IC-7000, while Kenwood – always innovative – offers the smallest and lightest *200-watt* transceiver around, the TS-480HX. That extra output power has found favour with travellers going to more remote locations who, for whatever reason, are unable to take a linear amplifier with them. The American company Elecraft has entered the fray and its K3 transceiver launched in 2007 won immediate plaudits from many purists, especially CW operators, in particular for its receiver specifications.

It was not only transceivers that were getting smaller and lighter. While linear power supply units (PSUs) using conventional transformers are preferred by many for base station applications, they can be very heavy and thus inconvenient for DXpedition use. The development of the switch-mode power supply (SMPSU) overcame this

PHOTO: 9M6DXX

The Watson 'Power-Mite-NF' switch-mode power supply provides 13.8V at 25A peak in a package weighing just 1.2kg.

objection and they too have become smaller and lighter over the years. Now, an SMPSU rated at 22A continuous and 25A peak, such as the British-badged Watson 'Power-Mite-NF', can weigh as little as 1.2kg and be smaller than the tiniest of rigs.

You could be forgiven for thinking that since the laws of physics have not changed over the years, HF antennas were always going to be big and heavy too. However, we should all be grateful to those amateurs who have developed innovative ways of providing multi-band operation from a single antenna or designed lightweight beams from wire and fibre-glass spreaders where once aluminium booms and elements were used. The result is that DXpeditions that once could not consider taking a typical 3-element triband aluminium Yagi because of its weight are now using antennas such as the 'Spiderbeam' or 'Hexbeam'.

So HF DXpeditioning, and in particular the so-called 'one-man DXpedition', has never been easier, nor so popular. It was not always like this though . . .

DXPEDITIONING PIONEERS

The 'father of the modern DXpedition' is generally acknowledged to be the late Bob Denniston, W0DX/VP2VI, who in 1948 was leader of the VP7NG 'Gon Waki' DXpedition to the Bahamas. ('Gon Waki' was a humorous homage to Thor Heyerdahl's *Kon Tiki* expedition of the previous year.) Denniston later operated from Clipperton Island as FO8AJ in 1954 and as HK0TU from Malpelo Island in 1969, both 'new ones' for DXCC. After he retired from business, he settled on the Caribbean island of Tortola, owning and operating the Smugglers Cove Inn, where he became a respected and well-known, if slightly eccentric, character in the British Virgin Islands.

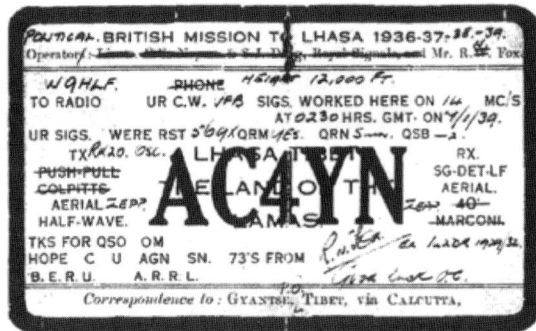

Was this the first DXpedition QSL card? AC4YN, originally operated by Sir Evan Nepean in 1936 and later by Sidney Dagg and Reg Fox.

An even earlier claim as the first DXpeditioner could be made on behalf of Sir Evan Nepean, Bt, G5YN, an English aristocrat who was the wireless officer on the British Political Mission to Tibet in 1936 and 1937. Employed to provide official communications between the British in Tibet and their headquarters in India, Sir Evan took the opportunity of operating on the amateur bands as AC4YN, providing a 'new one' for many of those early DXers. Later, the same callsign was used by Sidney Dagg and Reg Fox on the amateur bands. As a schoolboy in the late 1960s I had the pleasure of meeting Sir Evan when he gave a presentation on his operations from Tibet to a meeting of the Royal Signals Amateur Radio Society at Blandford Forum in Dorset. Surely this sparked an interest in DX operating in that young short-wave listener? Sir Evan's story was retold by Roger Croston in an article in *RadCom* in 2002 [1].

In the 1950s and 60s, when 'portable' equipment consisted of separate, valved, transmitters and receivers that were physically large and heavy, a one-man DXpedition was a major logistical challenge. One of the early pioneers was Danny Weil, a watch-maker by profession from London. Like Bob Denniston before him, Weil was inspired by Thor Heyerdahl's *Kon Tiki* expedition, and so he decided to build his own ocean-going yacht. He departed England's shores in 1954 on board

PHOTO: 9M6DXX

The Smugglers Cove Inn on the island of Tortola, British Virgin Islands. The ladder protruding from the roof supported wire antennas. This photograph was taken in May 2002, less than a week after Bob Denniston, VP2VI, became a Silent Key.

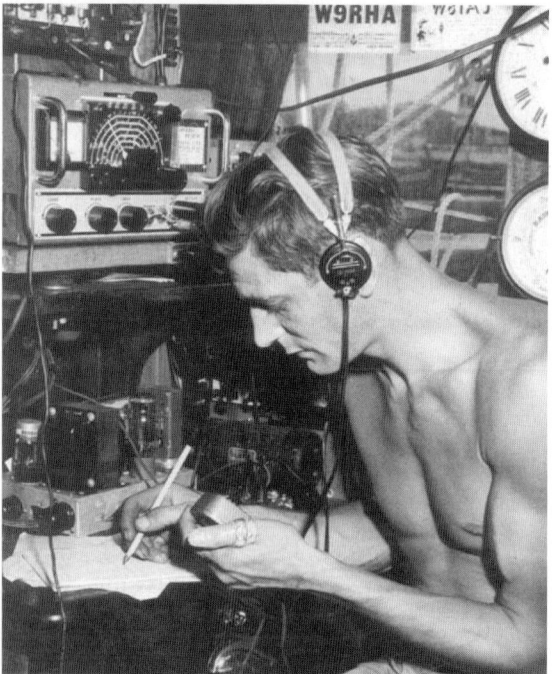

Danny Weil, operating as VP2VB/MM from on board Yasme, *circa 1959.*

the *Yasme*, intending to sail around the world.

Knowing nothing about amateur radio, Danny became friends with Dick Spenceley, KV4AA, who he met in the US Virgin Islands. Dick, who was a top DXer of the time, helped to teach Danny Morse code and encouraged him to take out his own amateur radio licence. This he did in the neighbouring British Virgin Islands, where he was licensed as VP2VB. Danny set up a station on board *Yasme* and then spent the next eight years travelling around firstly the Caribbean and later the Pacific, handing out well over 100,000 QSOs to the 'deserving' of the late 1950s and early 1960s.

Weil eventually achieved his goal of circumnavigating the globe in 1963, surviving shipwrecks and having to rebuild or replace *Yasme* twice. Once he achieved his goal, he gave up both sailing and amateur radio and eventually settled in the USA.

During the 1960s, Gus Browning, W4BPD, and Dr Don Miller, W9WNV, continued the tradition by activating a number of very rare and 'all-time new' countries.

Gus Browning, W4BPD, was first licensed as far back as 1925. In 1967 he was inducted as the first member of the *CQ* DX Hall of Fame. By the end of that year, Gus had made some 380,000 QSOs from no fewer than 117 countries. He was certainly the first amateur to have operated from

more than 100 DXCC entities, and there cannot be many who have repeated that remarkable feat since. It is all the more amazing when you consider both the difficulty of licensing - reciprocal licensing was still only in its infancy in the 1960s - and the cumbersome nature of the equipment in those days.

In the photograph below, Don Miller, W9WNV (second from left), is shown on board the New Zealand registered 30ft trimaran *Edward Bear* in December 1967. On this occasion he was setting off for Nelsons Island in the Chagos Archipelago, which he hoped would be counted as a new DXCC 'country' (as they were still called in those days). He claimed that Nelsons Island remained a dependency of Mauritius when the remainder of the archipelago became a part of the British Indian Ocean Territory along with the much larger island of Diego Garcia. Don operated as VQ8CBN but Nelsons Island never was added to the DXCC list.

It is said of Don's operating ability that he could identify up to 10 callsigns calling simultaneously in a CW pile-up and then remember them all, working each one in turn.

In 1980 Don was convicted of conspiring to murder his estranged wife. Although in the event no-one was harmed, Don nevertheless served a lengthy term in prison. He was released in 2002 and now holds the callsign AE6IY. A note on his qrz.com page reads, "Formerly W9WNV. Have finally returned to DXing. Contacts and inquiries welcome. Stay tuned and happy

Erik Sjölund, SM0AGD, seen here with a Butternut HF6V vertical antenna at 3C1AG, Equatorial Guinea, November 1989.

DXing from Don." Perhaps a whole new generation of DXers will yet witness a Don Miller, AE6IY, DXpedition?

In the late 1970s the advent of single-box transceivers and transistorised equipment led to a greater number of DXpeditions by such people as the late Lloyd and Iris Colvin, W6KG and W6QL; Erik Sjölund, SM0AGD (above), Martti Laine, OH2BH (below); and Jim Smith, originally G3HSR, later P29JS and now VK9NS. In the 1980s and 90s, DXpeditioning really took off and now, with tiny modern transceivers and a general relaxation in licensing in many countries, it has never been easier to put on a DXpedition.

Martti Laine, OH2BH (second from left), has been instrumental in opening up amateur radio in a number of countries, not least Albania. In 2003 he returned to the country with an IARU delegation to promote the hobby.

Rockall, which counts as EU-189 for the RSGB IOTA programme, is Europe's rarest IOTA island. As of mid 2008, it had still only been legally activated once.

DO-IT-YOURSELF DXPEDITIONING

There are two main types of DXpedition: the one- or two-man DXpedition, usually a 'holiday' operation, and the major multi-operator, multi-transmitter DXpedition. The only real difference is the amount of planning and organisation involved but we will discuss only relatively small-scale DXpeditions in any detail here. The book, *DXpeditioning Behind the Scenes* [2] - written by members of the 'Five Star DXers Association' and edited by Neville Cheadle, G3NUG, and Steve Telenius-Lowe, G4JVG / 9M6DXX - covers the planning, logistics and execution of a major DXpedition in great detail.

Two operators taking a single transmitter means that one station can be kept on the air for much longer periods by sharing the operating time and sometimes modes between the two operators.

UK operators are fortunate in having three quite rare DXCC entities on their doorstep which make excellent 'targets' for the first-time DXpeditioner. These are Jersey, MJ; Guernsey, MU; and the Isle of Man, MD. Close at hand too are numerous destinations in holiday companies' brochures, many of which are attractive DXpedition destinations, such as Crete, SV9, and Malta, 9H. Further afield, but still package holiday destinations and easily activated, are such places as Barbados, 8P; the Gambia, C5; and the Maldives, 8Q.

The RSGB Islands on the Air (IOTA) programme [3] is becoming more popular year on year, and this provides numerous island destinations, near and far, for the would-be DXpeditioner. In the British Isles alone, there are 28 IOTA island groups, ranging in difficulty of activation from the Isle of Wight to Rockall (above), which has only once been legally activated (a serious and well-planned attempt to activate it a second time in the spring of 2008 failed).

EQUIPMENT

Most single-operator DXpeditioners will choose the smallest and lightest transceivers available. At the time of writing (mid 2008) the single-operator DXpeditioners' rigs of choice are the Yaesu FT-857, Icom IC-706MkIIG or IC-7000, Kenwood TS-480SAT or HX, and Elecraft K2/100 or K3.

In the case of a multi-operator DXpedition, when two or more stations will be operating simultaneously, it pays dividends to check the transceivers' performance in the presence of strong out-of-band signals. A rig which may appear to be perfectly satisfactory (even on 40m, at night, using a 40m beam, during the *CQ* World Wide DX contest - surely the ultimate test of any receiver's front-end performance!) may still 'fall over' if you try to operate a second station on another band from the same location.

If you are planning to use two or more stations simultaneously (and this applies to non-DXpedition special-event stations too) consider investing in a set of band-pass filters such as the ICE or Dunestar [4] range. When the appropriate filter is inserted in the RF lead it will provide up to 40dB additional rejection of out-of-band signals, both on receive and transmit. If one is used on each station, a combined 80dB rejection of signals between the two bands can be obtained. Ensure the filters used are rated for at least 150W and, if you are using a linear amplifier, insert the filter *between* the driver and amp!

The only sure way to tell whether inter-station interference will be a problem is to try out all the rigs, in as close a simulation as possible, before departure. The Five Star DXers Association 9M0C Spratly, D68C Comoros and 3B9C Rodrigues DXpeditions used Yaesu FT-1000MPs, while the 2007 3B7C St Brandon expedition used FT-2000s, all with no problems at all, and members of the Voodoo Contest Group, which has operated as 5V7A from Togo, 9G5AA from Ghana and 3X5A

Richard Allisette, GU4CHY, points out the band-pass filter line-up at 7Q7MM, Malawi, April 2004.

from Guinea, among other locations, have recommended Kenwood TS-930S and TS-570D transceivers in this regard. More recently they have also successfully used Elecraft K2/100 transceivers.

If you are taking your valuable amateur radio equipment on an aircraft, you will probably want to carry it yourself and not entrust it to the baggage handlers. A small, lightweight transceiver, switch-mode power supply, microphone and/or key, headphones, and even a notebook PC for logging, plus the PC's PSU, can easily be accommodated in a single aircraft carry-on bag. Do be aware, however, that in addition to the maximum size requirement, airlines limit the weight of carry-on bags. Restrictions vary with some airlines setting a maximum weight limit of only 5kg for carry-on bags. A few are somewhat more generous and allow 6kg or sometimes even 10kg. Airlines are, of course, perfectly within their rights to weigh these bags at the check-in and to make excess baggage charges if the allowance is exceeded.

In recent years airlines in Europe have become much more strict about enforcing these weight limits and some (particularly the low-cost carriers) now no longer allow passengers travelling together to pool their allowances. In other words, if an airline allows a 15kg checked-in luggage allowance and one person checks in a 10kg bag, the other person travelling with them still only gets a 15kg allowance: they are not allowed a combined allowance of 30kg. In Asia, however, some budget carriers do still allow the pooling of allowances and although they can be strict in enforcing the checked-in weight limit they do not usually weigh carry-on bags.

AMPLIFIERS

It is important to be 'loud' when on a DXpedition, so you should consider taking a linear amplifier. Before rejecting the idea out of hand because of the weight, do consider how useful a linear might be. If your destination is a long-haul one to an area without a large concentration of relatively local amateurs to work, you may find the extra power almost essential. Locations such as the Maldives, 8Q; the Gambia, C5, or any Pacific island are fairly remote places where 100W to a simple wire or vertical antenna may prove disappointing. Also, if you plan to use the low bands you may find the additional power makes all the difference between not being heard at all on 160 and 80m and having a nightly DX pile-up.

The problem, of course, *is* the weight. Most HF linear amplifiers are rated at around 1kW output as this is the maximum power permitted in most countries (some allow 1.5kW and, especially in the Americas, there are a few

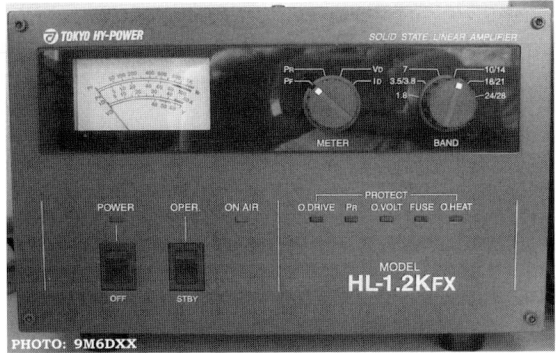

PHOTO: 9M6DXX

The Tokyo Hy-Power HL-1.2Kfx linear amplifier: 750W out in a (relatively) small box weighing a (relatively) light 15kg.

countries that permit 2kW output; the UK and former British territories are in the minority in only allowing 400W output). Generally these 1kW output amplifiers are too big and heavy to consider taking on a DXpedition. There are some, though, that should be considered: the SPE Expert 1K-FA is marketed as the smallest and most light weight 1kW amplifier available (it weighs 20kg). The Tokyo Hy-Power (THP) HL-1.5Kfx has similar specifications. The THP HL-1.2Kfx weighs less, at 15kg, and puts out 750W. THP also offers a 1.5kW solid-state PA, but this is a hefty 26kg. All three THP models have conventional linear PSUs built in. However, in the spring of 2008 THP announced that a 600W model, to be known as the HL-1.1Kfx, will shortly become available that will have a switch-mode power supply and weigh only 10kg. This will undoubtedly become a popular choice for weight-conscious DXpeditioners!

Many years ago now Steve Webb, G3TPW, built and marketed a small amplifier that he called the 'SRW Kilowatt Loudenboomer'. Using four PL519 TV sweep tubes, which have the advantage of being very cheap, it produces up to 400W PEP output in a package measuring 14W x 11D x 5Hin and weighing just 7kg (including built-in mains PSU). Loudenboomers are now difficult to obtain, but are ideal for a one-man DXpedition requiring more than 100W of power.

PHOTO: 9M6DXX

The 'SRW Kilowatt Loudenboomer' amplifier using TV sweep tubes. This weighs just 7kg but puts out just under 400W.

The Icom IC-2KL (despite its name, capable of about 500W output) is now also quite an old piece of equipment and can sometimes be found second-hand at a reasonable price. It comes in two small boxes: the transistorised RF deck which, at under 7kg, is light enough to be hand-carried, and a separate linear PSU weighing 13.9kg, which could probably be packed in a Samsonite-type suitcase, if surrounded by plenty of bubble wrap and holiday T-shirts, and sent as checked-in luggage.

There are numerous transistorised linear amplifiers now on the market that were originally designed for 11m CB use, but which have been modified for the HF amateur bands. Most require separate high current (e.g. 40A - 50A) 12V PSUs but an exception is the Sommerkamp SLB-300, a 220V AC-operated amplifier capable of 265W RMS output and sold in the UK for under GBP £400.

Even if a lightweight solid-state amplifier is not available, with two or more members of a DXpedition team it could still be worth considering taking a conventional valve (US: tube) amplifier. They weigh in from 14.5kg (for the Ameritron AL-811X), though most are around 25kg or more. If packed well, preferably in the original box, they should survive the rigours of an aircraft journey. Glass valves should of course be removed and hand-carried, wrapped in plenty of bubble wrap, although manufacturers of ceramic tube amplifiers often recommend leaving the valve in place during shipping of their amplifiers (check with the manufacturer concerned).

ANTENNAS

Arguably the most popular antenna for the one-man DXpedition is the Butternut HF6V (right). This is a six-band vertical, 7.95m (26ft) high, covering 80, 40, 30, 20, 15 and 10m (but not 12 and 17m) and which uses resonating circuits rather than traps. It requires radials in order to work properly. This antenna is popular with DXpeditioners as it packs into a box measuring just over 1m (40in) long by 12cm (4in) square and weighing 4.5kg, and is thus easy to check-in on commercial flights. It performs well, especially so if the DXpedition destination is an ocean-side location: indeed recent experience on DXpeditions has shown that in terms of low-angle gain a simple quarter-wave vertical can work as well as a 4-element Yagi *if* mounted very close to the ocean.

Other verticals, such as the Cushcraft R8 and R6000, also work well. They have two advantages over the Butternut: 12 and 17m are included (although 80m is not included on the R8, and 30,

40 and 80m are not on the R6000) and, being half-wave design antennas, they do not need long wire radials (short 'spoke' radials are part of the design). Their disadvantages are that they do not pack into such a small box and that they are somewhat heavier, as they use traps.

Single-band dipoles work well too, and have the advantage of being extremely lightweight and easy to transport. Most DXpeditions, even single-man ones, will want to operate on several bands though, and, depending on the location, it may prove difficult to put up several different single-band antennas. Multiband wire antennas, such as the G5RV or Carolina Windom, tend not to be as popular as multiband verticals on DXpeditions. In most cases an external ATU will be required, making it necessary to transport one more box plus an additional coax patch lead.

For multi-operator DXpeditions, a triband Yagi should be considered. An antenna such as the Cushcraft A3S weighs under 13kg and so is feasible to transport if there are two or three operators and they are able to pool their weight allowances. There's no question that, at most locations, a beam will make a real difference to the final QSO count. Probably more difficult than transporting a beam, though, is taking something to mount it upon. Telescopic 'push-up' masts are ideal, though they can be as bulky and heavy to transport as the antenna itself. The minimum height for a beam to perform reasonably well on the 20m band is about 10m (around 30ft). If the DXpedition is operating from a high-rise hotel and you can gain access to the roof, beams mounted on short poles of 2 - 3m should work well, providing the antenna is mounted close to the edge and 'looking over' the roof. Here, though, you will probably also need a rotator, an additional complication and weight to consider.

LOGGING

These days almost all DXpeditions log by computer. The exception is the single-operator DXpedition on a 'holiday' type operation which does not expect to make a very large number of QSOs. If you are using paper, a standard log book will probably be the method of choice, although if you are really operating in 'DXpedition mode' and working a pile-up quickly you will probably not be stopping for a chat and will be giving 59 or 599 reports to allcomers. If that's the case you could log in an exercise book, with the date, band and mode at the top of the page, and the list of callsigns worked in a series of columns down the page. If the pile-up is thick and fast it is only necessary to note the time every five or 10 QSOs, i.e. every two or three minutes. The well-known German DXpeditioner Baldur Drobnica, DJ6SI, used this method in the days before computer logging.

There's no doubt, though, that computer logging makes log checking and QSLing much quicker and easier. Most of the popular station logging programs work fine for DXpeditions. Some contest logging programs, including K1EA's *CT* [5], also have a 'DXpedition mode' which allows you also to log QSOs made on 10, 18 and 24MHz, for example. *CT* is now available as freeware and may be downloaded from the Internet.

CUSTOMS

Customs regulations vary from country to country and it is beyond the scope of this book to go into detail. There are no customs formalities between European Union (EU) countries, so 'holiday DXpeditions' to any EU destination should not present any difficulties. These countries have a 'Blue Channel' in addition to the more familiar Red and Green customs channels for travellers from within the EU.

Most 'tourist-friendly' countries will permit the temporary import of a limited amount of amateur radio equipment as part of the traveller's personal belongings. There is a limit, however, and while a single, obviously used, transceiver, power supply and multi-band vertical is likely to be considered personal belongings by all but the sternest of customs officials, several boxed new transceivers, linear amplifiers, beams, masts and 100m drums of coaxial cable turning up without the necessary importation documents will certainly not be so considered. If you are planning a multi-operator DXpedition with that sort of equipment count, do your homework beforehand and find out what paperwork is required.

Often a *carnet* is the simplest: all the equipment must be listed in an inventory and it must be re-exported to the country of origin. It's important to note that such paperwork does not come cheap. In a major multi-operator DXpedition the cost of customs clearances and paperwork can amount to almost as much as the freight charges. This subject is covered in detail in reference [2].

REFERENCES

[1] 'The story of AC4YN - a radio adventure in Tibet, 1936', Roger Croston, *RadCom* June 2002.
[2] *DXpeditioning Behind the Scenes*, eds Neville Cheadle, G3NUG, and Steve Telenius-Lowe, G4JVG, Radio Active Publications, 2000.
[3] *RSGB IOTA Directory*, edited by Steve Telenius-Lowe, 9M6DXX, RSGB, 2007.
[4] Dunestar Systems, PO Box 37, St Helens, OR 97051, USA; website: www.dunestar.com; e-mail: dunestar@QTH.com
[5] *CT* version 10, K1EA Software; website: www.k1ea.com

All you wanted to know about licensing (but were afraid to ask)

Before you can operate from overseas - whether it be a simple holiday operation or a major DXpedition - you must, of course, have a licence. That goes without saying. But depending on your chosen destination, obtaining a licence can be anything from a mere formality to one of the most difficult steps in the whole operation.

At one extreme is the so-called 'CEPT Licence' with which most readers, whether from Europe or North America, will already be familiar. With the inclusion in 1999 of the USA and its many overseas territories it has become possible to operate from over 100 DXCC entities and numerous IOTA islands without the necessity of having to apply for overseas licences.

At the other end of the scale would be an attempt to operate from the Democratic People's Republic of Korea (North Korea), P5, or any of those other countries where it is normally considered 'impossible' to get a licence. Most countries, of course, fall somewhere between, on a long sliding scale of difficulty.

THE CEPT LICENCE

So, just what is the CEPT Licence? CEPT stands for the European Conference of Postal and Telecommunication Administrations (in French) and it is their Recommendation T/R 61-01 which allows radio amateurs from countries which have implemented the recommendation to operate from any of the other countries, with the minimum of formalities. This Recommendation T/R 61-01 is usually referred to, unofficially, as 'The CEPT Licence'.

Many European countries, east and west, have now implemented T/R 61-01, and a number of countries outside the CEPT area have also agreed to allow amateur radio operation under the terms of the CEPT Licence. The most recent of these to

The Caribbean island of St Barts: a new DXCC entity and covered by the CEPT Licence.

PHOTO: 9M6DXX

Country	Prefix
Australia	VK
Austria	OE
Belgium	ON
Bosnia-Herzegovina	E7
Bulgaria	LZ
Canada	VE, VO, VY
Croatia	9A
Cyprus	5B
Czech Republic	OK
Denmark	OZ
Estonia	ES
Finland	OH
France	F
Germany	DL
Greece	SV, SW
Hungary	HA, HG
Iceland	TF
Ireland	EI, EJ
Israel	4X, 4Z
Italy	I
Latvia	YL
Liechtenstein	HB0
Lithuania	LY
Luxembourg	LX
Macedonia	Z3
Monaco	3A
Netherlands	PA
Netherlands Antilles	PJ
New Zealand	ZL
Norway	LA
Peru	OA
Poland	SP
Portugal	CT
Romania	YO
Slovakia	OM
Slovenia	S5
South Africa	ZS
Spain	EA, EB
Sweden	SM, SA
Switzerland	HB9
Turkey	TA
Ukraine	UT
United Kingdom	M
USA	W

Table 1: List of countries that have implemented CEPT Recommendation T/R 61-01 (as of 1 June 2008). Read CEPT Recommendation T/R 61-01 carefully to determine the conditions under which you may operate from a specific country, and for the precise prefix that should be used with your callsign.

join was Australia, in May 2008, and the others are Canada, Israel, the Netherlands Antilles, New Zealand, Peru, South Africa and the USA.

The CEPT Licence thus covers the majority of locations for 'holiday' operations, the exceptions being long-haul (from Europe) destinations such as the Gambia, the Maldives, Malaysia, and most of the Caribbean islands (although the Netherlands Antilles, the French Caribbean islands of Martinique, Guadeloupe, St Martin and St Barts, and the US Virgin Islands and Puerto Rico, *are* covered by the CEPT Licence).

There are some important points to note about CEPT Licence operation:
• T/R 61-01 allows for temporary portable operation only. This is generally regarded as meaning up to three months at a time (and 'Portable' in this context includes operation from buildings such as hotels or from other licensed amateurs' stations.)
• Although you do not have to go through the procedure of applying for a local licence, it is the responsibility of the individual amateur to find out the frequency bands permitted, the appropriate band limits, the maximum power permitted, whether 6m operation is allowed, etc, if necessary by contacting the appropriate licensing authority. It must not be assumed that the licensing conditions are the same as those of your home licence, for very frequently they are not. Because this information changes frequently, I know of no single database containing up to date details. In most cases, however, the national amateur radio society in the country concerned will be able to help you.
• You must take your current home licence with you when operating abroad under the terms of the CEPT Licence (and, in the case of the new UK 'lifetime' licence, the whole document, not just the first page).

It sometimes appears to be almost impossible to keep up with the countries that have implemented T/R 61-01. In the UK, even the various licensing authorities over the years have on occasions not received official notification of a country having implemented the CEPT Licence and so have not advised UK amateurs of the fact, sometimes even many months after the event.

The *only* fully definitive listing of countries that have implemented CEPT Recommendation T/R 61-01 can be found on the European Radiocommunications Office (ERO) website, by going to www.erodocdb.dk and then clicking on T/R 61-01 under "ECC Recommendations". Then click on "Implementation" under "CEPT Radio Amateur Licence". The complete T/R 61-01 document is available for downloading by clicking on the *Microsoft Word* or *Adobe PDF* symbols at the top of the page. The list of countries that have implemented T/R 61-01, current as of 1 June 2008, is reproduced here as **Table 1**.

From a DXpedition point of view remember that there are also many *additional* DXCC entities associated with these countries and many of these can be attractive DXpedition locations. For example, Finland is a CEPT Licence country, and so the licence is also valid in the Åland Islands, OH0, and Market Reef, OJ0; Portugal includes the Azores, CU, and Madeira, CT3; and the USA covers many overseas territories including the US Virgin Islands, the Northern Marianas, American Samoa etc. These additional DXCC entities are shown in **Table 2** opposite.

Most people realise that if they want to operate from a British overseas territory, such as Turks and Caicos, VP5, or the Falkland

DXCC Entity	Prefix	DXCC Entity	Prefix
Åland Islands	OH0	Kure Island*	KH7
Alaska	KL7	Lord Howe Island	VK9
American Samoa	KH8	Madeira	CT
Auckland & Campbell Islands*	ZL9	Market Reef*	OJ0
Azores	CU	Martinique	FM
Baker & Howland Islands*	KH1	Mayotte	FH
Balearic Islands	EA6	Mellish Reef*	VK9
Canary Islands	EA8	Midway Island	KH4
Ceuta & Melilla	EA9	Mt Athos*	SV/A
Chatham Islands	ZL7	Navassa Island*	KP1
Christmas Island	VK9	New Caledonia	FK
Clipperton*	FO	Norfolk Island	VK9
Cocos (Keeling)	VK9	Northern Ireland	MI
Corsica	TK	Northern Mariana Islands	KH0
Crete	SV9	Palmyra, Jarvis Islands*	KH5
Crozet Island*	FT	Prince Edward & Marion Islands*	ZS8
Desecheo Island*	KP5	Puerto Rico	KP4
Dodecanese Islands	SV5	Réunion	FR
Faroe Islands	OY	Sable Island*	CY
French Antarctica (Terre Adelie)*	FT	Sardinia	IS0
French Guyana	FY	Scotland	MM
French Polynesia	FO	Sint Maarten, Saba & St Eustatius	PJ
Glorieuses*	FR	St Barthelemy (St Barts)	FJ
Greenland	OX	St Martin	FS
Guadeloupe	FG	St Paul & Amsterdam*	FT
Guam	KH2	St Paul Island*	CY
Guernsey	MU	St Pierre & Miquelon	FP
Hawaii	KH6	Svalbard	JW
Isle of Man	MD	Tromelin*	FR
Jersey	MJ	US Virgin Islands	KP2
Johnston Island*	KH3	Wake Islands*	KH9
Juan de Nova*	FR	Wales	MW
Kerguelen*	FT	Wallis & Futuna	FW
Kermadec Islands*	ZL8	Willis Island*	VK9
Kingman Reef*	KH5K		

Table 2: List of *additional* DXCC entities that are covered by the CEPT Licence. Note that the CEPT Licence covers *only* amateur radio licensing: it is necessary to obtain separate permission to visit, land or operate on many of these territories. Those entities where it is *known* that landing permission is essential or recommended are marked with an asterisk (*).

Islands, VP8, they have to obtain a local licence. Such is not the case with French overseas territories, however. Whilst it is well known that France has implemented T/R 61-01, it is perhaps not so widely realised that you can operate from the French Caribbean islands, or Mayotte in the Indian Ocean, for example, under the terms of the CEPT Licence. A recent change to the definitive listing of CEPT Licence countries on the ERO website is that it now includes *all* the French overseas territories. Until recently it was necessary to obtain local licences for operation from some French outposts, including New Caledonia and French Polynesia, but this is apparently no longer the case (although whether the local PTT officials in Noumea and Papeete

agree with this decision - or are even aware of it - is unclear). However, although *licensing* is covered by CEPT it *is* still necessary to obtain local permission to visit and/ or operate from certain French overseas territories, in particular the Indian Ocean dependencies administered from Réunion (Iles Glorieuses or Glorioso, Juan de Nova, Tromelin and Europa Island).

In most countries the use of a call district number by visiting CEPT amateurs is not obligatory. However, whether mandatory or not, if the call district system exists its use is always to be recommended, as the information is of use to your fellow amateur. For example, although the use of the call district indicator is not obligatory in Spain, other amateurs would be interested to

PHOTO: 9M6DXX

The Cocos (Keeling) Islands, VK9C: one of the many additional DXCC entities included in the CEPT Licence (see Table 2 on page 17).

know whether you are operating from the Spanish mainland, the Balearic Islands, EA6; the Canary Islands, EA8; or Ceuta and Melilla (Spanish North Africa), EA9, as these count as four different DXCC entities. In Finland, use of the call district number is not obligatory, except when in the Åland Islands, when OH0 should be used as the prefix. In Canada you should use VO as the prefix for Newfoundland and Labrador, VY for the Yukon Territory and Province of Prince Edward Island, and VE for the remainder of Canada. Once again, full details are in the various Appendices to the complete T/R 61-01 document published on the ERO website.

Another anomaly: most countries require that you use their callsign prefix before your own callsign, e.g. F/M0QQQ from France, but Australia (VK) and Peru (OA) require that you sign your own callsign first, appending the country prefix *after* your own call, e.g. M0QQQ/VK. Peru also requires a call district number, e.g. M0QQQ/OA4.

So much for the basics of the CEPT Licence. Now, unfortunately, it gets a little more complicated, particularly if you are either a relatively new licensee or an American General or Technician class licensee.

Following the ITU World Radio Conference 2003, in which the international requirement for a Morse code qualification for use of the HF bands below 30MHz was removed, CEPT Recommendation T/R 61-01 was revised to reflect this change. However, because some countries that had previously implemented T/R 61-01 decided that, international regulations notwithstanding,

they would *keep* a Morse code test for HF licences, they have not adopted the revised Recommendation. This means that some countries insist that you should have passed a Morse code test in order to operate from their territory on the HF bands under the terms of the CEPT Licence. (The fact that you are able to operate on the HF bands at home is irrelevant.) The point to remember is that the terms and conditions under which you may operate are set by the country you are visiting and *not* by your home licence. What this means in practice is that if you were licensed *after* 2003, or if your licence was a VHF-only one *prior to* 2003, you *may* either be restricted to frequencies of 50MHz (or 144MHz) and higher, or be required to take a Morse code test in order to operate on HF.

The situation changes frequently, so refer to the ERO website for the latest information.

CEPT Recommendation T/R 61-01 was amended once again in 2008, this time affecting many US amateurs. On 4 February 2008, CEPT revised its table of equivalence between American amateur licence classes and the CEPT licence. Full CEPT privileges are now only granted to Americans with Amateur Extra or Advanced class licences. American General or Technician licensees are no longer valid under the CEPT Licence scheme (and the US Novice class licence never has had CEPT Licence equivalence).

The IARU Region 1 website at www.iaru-r1.org has some useful advice for radio amateurs intending to operate under the terms of CEPT Recommendation T/R 61-01. Click on "Operating abroad".

HAREC

An interesting extension of the CEPT Licence is the HAREC licence. This is the 'Harmonised Amateur Radio Examination Certificate', which was launched in the UK in July 1996. The HAREC is now accepted by many countries as a qualification for a permanent full (not a temporary reciprocal) licence, thus avoiding the necessity of having to sit the local examinations. A normal callsign is issued, e.g. an overseas amateur using a HAREC as a qualification would be granted a full M0 callsign in the UK.

The definitive list of countries that have implemented CEPT Recommendation T/R 61-02 (the HAREC certificate) can be found on the European Radiocommunications Office website, by going to www.erodocdb.dk and then clicking on T/R 61-02 under "ECC Recommendations". Then click on "Implementation" under "Harmonised amateur radio examination certificates". The complete T/R 61-02 document is available for downloading by clicking on the *Microsoft Word* or *Adobe PDF* symbols at the top of the page.

The HAREC is intended to be used by licensees who are either moving permanently to another country, or visiting that country on a regular basis, particularly for extended periods (e.g. greater than about three months). Some countries may require the holder of a HAREC also to be in possession of a residence visa before they will issue a full licence. You will need a permanent mailing address in the country concerned, although this could be that of a radio amateur friend whom you visit when in that country and whose station you use when you are there.

IARP

Although accepted by several countries outside the region, the CEPT Licence and the HAREC are both European initiatives. In the Americas a similar scheme called the International Amateur Radio Permit (IARP) has been in operation since June 1995, although it still covers relatively few countries.

The IARP is an initiative of CITEL (*Comision Interamericana de Telecomunicaciones*), the Inter-American Telecommunication Commission, which in turn is an entity of the Organisation of American States (OAS).

The IARP allows radio amateurs from participating states to operate from the other participating states without having to apply for a licence in the country concerned. However, unlike the CEPT Licence which is itself the required document, it is necessary to apply for and obtain a separate IARP in your home country before you operate from other countries.

There are still two classes of IARP: Class 1 with a Morse code proficiency requirement and covering all bands, and Class 2 with no Morse requirement but covering amateur bands above 30MHz only.

IARP participating countries are: Argentina, Brazil, Canada, El Salvador, Panama, Peru, Trinidad and Tobago, United States of America, Uruguay and Venezuela.

The full text of the IARP legislation can be found on the OAS CITEL website at www.citel.oas.org/iarp.asp

Amateurs from the USA wishing to apply for an IARP may do so by going to www.arrl.org/FandES/field/regulations/io/#IARP and downloading an application form.

RECIPROCAL AND UNILATERAL

The CEPT Licence and IARP are all very well, but what if you want to operate from, say, Ghana, Anguilla, Fiji or Malaysia? This group of countries - and numerous others - will grant a licence to nationals of most countries on the basis of the amateur radio licence issued in their own country.

The term usually used is Reciprocal Licence, although this is something of a misnomer, for in many cases there is no *reciprocal* agreement at all; rather it is a unilateral arrangement. This is the case, for example, in Anguilla, VP2E: a British amateur can obtain a licence in Anguilla, but there is no official reciprocal licensing agreement, and an Anguillan cannot necessarily obtain a UK call on the basis of his Anguillan licence.

A number of countries have reciprocal licensing agreements with a restricted number of countries only. For example, a British, Australian, German or American amateur can obtain a reciprocal licence in Papua New Guinea, P2, but not a Japanese amateur (in fact at least three Japanese amateurs have operated from P2, but they have had to sit the local licence examinations). Japan only has reciprocal licensing agreements with nine countries: Australia, Canada, Finland, France, Germany, Ireland, Korea, Peru and USA (but not the UK).

There is yet another group of countries which have reciprocal licensing agreements with certain countries, but which will normally only issue licences to those with residence permits, e.g. expatriate workers, but not to short-term visitors such as tourists - or DXpeditioners. This group includes Singapore, 9V; the United Arab Emirates, A6, and Saudi Arabia, HZ. See the Licensing Information section of this book for further details. Until a few years ago Kenya, 5Z, was also among these countries, but their licensing regime was liberalised around 2004.

In some countries there may only be one individual who can issue the licence; sometimes this is the Minister of Telecommunications himself. If this is the case, even if you have received written confirmation from that individual that a licence will be issued to you after you arrive in the country, you should ensure that he will be available to meet you when you do arrive: the Minister could be attending a session of parliament, he could be on a political rally up-country, or he could even be on a two-week family holiday in Florida! Here the importance of local

It was not always impossible to obtain a licence in Yemen: 7O1AA was operated by three Kuwaiti amateurs in June 1990.

knowledge cannot be overstated. In many small countries the local radio amateurs know the licensing officer(s) on a personal basis and may well be able to obtain a licence on your behalf where you would fail.

Finally, some countries do not normally issue licences at all, the best-known example being the Democratic People's Republic of Korea (North Korea), P5, although there are others, such as Yemen, 7O, and Tunisia, 3V, where it is still very difficult, if not actually impossible, for an individual to obtain a licence. (Tunisia has allowed the operation of club stations, but they do not normally issue licences to individuals.)

However, the pendulum may swing one way and open up one country for a while, then it swings away again and as that one closes, so another will be opened up. Many DXpeditioners keep a weather eye on changes in world politics which can suddenly create an 'opening' and allow them to obtain a licence from a really rare spot. Some countries can take decades to open up - look

at China, BY; Vietnam, 3W, or Albania, ZA. Rwanda, 9X, did not issue any amateur radio licences for around 10 years, but a change of personnel in the national PTT in 2007 has led to a more liberal licensing policy. Most countries that at present are considered either very difficult or virtually impossible *have* issued licences - even to foreigners - in the past, for example Yemen, 7O, and Tunisia, 3V, and undoubtedly they will again at some time in the future.

Even if you want to operate from one of the 'difficult' countries which will not normally grant a licence, all is not necessarily lost. Many countries allow (or at least turn a blind eye) to DXpedition-type activities from the station of a properly-licensed local operator or club station. There have been a number of operations from Egypt, SU; the Cape Verde Islands, D4; and the United Arab Emirates, A6, for example.

OBTAINING THE LICENCE

Although I am now resident in Malaysia and licensed here as 9M6DXX, I first operated from Malaysia as 9M2/G4JVG when on a holiday to Penang in April 1996. In those days, although licensing for long-term residents was not a problem, not very many Malaysian licences had been issued to short-term visitors, and I had been told by others who had previously operated from Malaysia that it might take nine months or more to obtain a licence. As my visit was a 'last-minute' holiday, booked through a travel agent only a month before departure, I therefore had less than a month to get the 9M2 licence.

The licence application procedure involved sending sketches of the antennas to be used, a copy of the circuit diagram of the transmitter and a sworn statement made in front of a Justice of the Peace, Magistrate or a Commissioner of Oaths that I would keep secret any communications accidentally overheard on the radio, only divulging them to the appropriate Malaysian authorities. All this was in addition to the usual application form with name, address, nationality, passport number, exact location of operation, frequency bands required, transmitter power etc, and photocopies of the home licence, passport personal details pages and so on. The eventual package was quite bulky and off it went by air mail to the licensing authority in Kuala Lumpur.

Two and a half weeks went by and I had heard nothing, so, because of the time difference, I rose before dawn on several occasions to make phone calls to them. After a number of attempts on different days I was eventually put through to the right person in the right department and was told that yes, they had my application on file, but because I was intending to operate from Penang I should have sent my application to their local office there instead. Was I stopping in Kuala Lumpur *en route*? If so, I could pick up my licence from them on the way. In fact I was only spending an hour at the airport in KL, so this was not possible. They told me to go to their office in Penang after I arrived and pick up my licence in person. They would forward on the papers for me. No, it was not possible to post the licence to England - and besides, there probably was not enough time remaining before my departure.

Sure enough, my licence was waiting for me at the regional licensing office in Penang, where I discovered that if I had just turned up with my home licence and passport they would have issued the licence for me there and then, over the counter, in about half an hour!

An important lesson was learned. In Malaysia, as in many countries, the authorities much prefer to issue an amateur radio licence to the individual in person. Presumably the reason for this is that they like to see just who is operating from their country (although I have never known anyone being refused a licence because they did not like the look of him!) My experience in many countries has been the same, from Malta, 9H; to Vanuatu, YJ; and Samoa, 5W (fortunately it is not the case in Fiji, where the licensing authority is in the capital, Suva, around 200km by road from the international port of entry at Nadi!)

THE MAN ON THE GROUND

Two years later I was one of the team members planning a DXpedition to the Malaysian island of Pulau Layang Layang in the Spratly Islands. As I had applied for, and obtained, a Malaysian licence less than two years previously, I was the one tasked with obtaining the licence for the DXpedition.

Although Layang Layang is geographically part of the Spratly Islands group it is considered by Malaysia to be sovereign territory and so there was no question of operating from there without first obtaining a valid Malaysian licence. Indeed, the ARRL DXCC desk has made it clear that they will not accept for DXCC credit any further operations from any island in the Spratly group which use 'self-assigned' callsigns, such as those beginning with the prefix 1S. Any Spratly operation must be properly licensed by whichever country is the occupying force on the island concerned.

We wanted a callsign which would be recognisably a Spratly Islands callsign - and in 1998 this meant a 9M0 call. Visitors operating in Malaysia are normally granted callsigns in the form of 9M2/, 9M6/ or 9M8/ own callsign. Layang Layang is administered from Sabah, but the licensing office there is only authorised to issue 9M6 - not 9M0 - callsigns. Although there were precedents of individuals operating from Pulau Layang Layang using 9M6 callsigns (and these operations were accepted for Spratly by the DXCC desk), we did not want there to be any confusion among 'casual' operators that our operation was from East Malaysia. The answer was to apply for a Special Group Licence from the licensing authority's headquarters in Kuala Lumpur.

This is where Ray Gerrard, G3NOM, comes into the story. In mid-1997, the time we were applying for the licence, Ray was living in Kuala Lumpur and was licensed as 9M2OM. I had already applied for a special 9M0 callsign and had made several early morning phone calls to Kuala Lumpur once again before Ray became involved.

Ray's diplomatic and negotiating skills were put to the test and after many visits in person to the licensing authority building he was able to secure the Special Group Callsign 9M0C for us and, most importantly, he received the licence document itself (see **Fig 1** on page 22) by hand. This illustrates another truth about licensing; *never* underestimate the importance of having a 'man on the ground' - a local liaison man who can sort things out. It is so much easier to do this in person than over a long-distance telephone link, or by fax or e-mail.

If you are going on a multi-operator DXpedition, check to determine whether each operator requires an individual licence, or if one group licence will cover everybody. If in doubt, everyone should apply for individual licences.

Layang Layang in the Spratly Islands.
(PHOTO: 9M6DXX)

JABATAN TELEKOMUNIKASI MALAYSIA
KEMENTERIAN TENAGA, TELEKOM DAN POS
(Department of Telecommunications)
(Ministry of Energy, Telecommunications and Post)
WISMA DAMANSARA
JALAN SEMANTAN
50668 KUALA LUMPUR

Telefon *(Telephone)* : 603-2556687
Kawat *(Cable)* : DIRGENTEL
Teleks *(Telex)* : GENTEL MA 28020
Fax : 603-2530508
E-mail : jtmhq@tm.net.my

Mr.Stephen Telenius-Lowe
c/o Mr. Ray Gerrard (9M2OM)
16 JalanBkt. Antarabangsa
Taman Bkt. Mewah
68000 Ampang.

Ruj. Tuan
Your Ref :

Ruj. Kami
Our Ref : JTM 10/7580.01(32)

Tarikh
Date : 4th. September 1997

Dear Sir,

SPECIAL GROUP CALL SIGN FOR AMATEUR RADIO
ACTIVITY AT PULAU LAYANG-LAYANG, SABAH, MALAYSIA.

We refer to the above matter.

2. Your application for the use of callsign **9M0C** has been approved, with the following conditions:-

i) the station shall be operated only at the the specified times and premise, as follows:-

Venue : Pulau Layang-layang
Sabah, Malaysia
Date : February - April 1998
Time : Please state the times when the programme starts and ends

Fig 1: The 'Special Group' permit for the use of 9M0C from Pulau Layang Layang in the Spratly Islands.

OTHER PERMITS

Sometimes receiving the amateur radio licence is only half the battle. There are many DXpedition locations around the world where it is perfectly possible to obtain a licence, but it is difficult or impossible to receive permission to visit the territory in question. Examples that come to mind include Navassa Island, KP1, and Desecheo Island, KP5, in the Caribbean where obtaining a US licence is a mere formality but where permission to visit the islands must be obtained from the US Department of the Interior. This is much harder to come by. Other examples include Kermadec, ZL8; Auckland and Campbell, ZL9; Mt Athos, SV/A; the Glorioso Islands, FR/G, and the other dependencies of Reunion, and indeed Pulau Layang Layang, where it is necessary to obtain permission from the Royal Malaysian Navy before operating from the island.

Then there are those territories where it may not strictly be obligatory to receive landing permission, but where it is wise to inform the authorities of one's intended operation and to receive their approval. Mellish Reef, VK9M, an uninhabited sand bar surrounded by sharp coral reefs in the Coral Sea east of Australia, is a case in point. Being uninhabited, and so far from civilisation, there is no-one to prevent you from landing there. Furthermore, an Australian licence is easy to come by, yet Mellish Reef is an important breeding ground for sea birds and to prevent difficulties for future DXpeditioners it is always good to ensure that the powers that be, who have the right to refuse permissions, are aware of your intentions. You may well be asked to sign a declaration that you will not interfere in any way with the wildlife on the island. Some DXpeditions to Kermadec, ZL8, and the Auckland and Campbell Islands, ZL9, were obliged to have wildlife officers travel with the group to supervise their operation and to ensure that no harm befell the environment as a result of the visiting DXpedition teams.

Another reef, this time in Europe, is a further example of an area which may not require landing permission, but where it is nevertheless wise to ensure the co-operation of the authorities. Market Reef, OJ0, in the Gulf of Bothnia between the Finnish Åland Islands, OH0, and the Swedish coast, is a DXCC entity and IOTA island which can be activated under the terms of the CEPT Licence. However, it is a tiny rock where, even in summer, waves can wash over the entire island. If only for reasons of personal safety it is therefore necessary to use the buildings on the island (an automated lighthouse and ancillary huts) for any amateur radio operation - and that means obtaining permission from the appropriate authorities and getting the keys to the buildings before setting out on the DXpedition.

Some DXCC entities are military outposts and it is vital to ensure the military have no objection to an amateur radio operation taking place from

The monastery of Simonos Petras on Agion Oros, or Mt Athos. A special permit is required to operate from Mt Athos, even if you have a licence valid in Greece.

'their' island. Do not assume that because you have a valid licence from the local PTT office you can necessarily operate from anywhere in that country. It is the responsibility of the would-be DXpeditioner to determine that either no permission is required, or, if it is, to obtain that permission. Never just go along and hope for the best: you may be refused operating permission after you have already arrived in the country, which would not only be disappointing to you, but an expensive waste of money. Even worse, you could put back by a decade or more a 'legitimate' DXpedition group which had been working away in the background for years to obtain permission through the correct routes.

Examples of DXCC entities which are rare because of difficulties in obtaining military permission are the Andaman Islands, VU4, and Lakshadweep (or Laccadive) Islands, VU7. Both island groups, which are separate DXCC entities, are politically part of India and both have tourist resorts on them. It is easy to travel there as a tourist and it's perfectly possible (though it can take a long time with considerable bureaucratic wrangling) for a foreign amateur to obtain an Indian licence. However, a separate permit is required for operation from either group of islands, and this is not normally granted, either to foreign visitors or even Indian nationals. Naturally, it is for this reason that the Andamans and Lakshadweeps became so rare. However, amateur radio in India received a boost and much good publicity following the work carried out by Bharathi Prasad, VU2RBI, and her team who coincidentally were on a DXpedition to the Andaman Islands at the time of the massive earthquake and tsunami in December 2004. Thanks to the excellent emergency communications work carried out by the Andaman DXpedition team it has since become easier for Indian and foreign amateurs alike to activate the Andaman and Lakshadweep Islands, although special permission *is* still required for both of these territories.

Then there are disputed territories such as the Western Sahara, the Turkish Federated State of Northern Cyprus and the Spratly Islands. If you want your DXpedition to count for DXCC you must take heed of what the ARRL DXCC desk has

Bharathi, VU2RBI, whose work while on a DXpedition to the Andaman Islands at the time of the Asian tsunami led to a relaxation of the licensing rules in India.

to say about such areas. We have already discussed licensing on the Spratly Islands and said that a 'self-assigned' callsign such as a 1S call will not be accepted for DXCC credit. Instead it is necessary to obtain a 'proper' licence from the licensing authority of whichever country holds *de facto* sovereignty on the particular island you operate from. Since this clarification of the rule was made by the DXCC desk, most Spratly operations have been from the Malaysian-administered island of Pulau Layang Layang, but there have been a few from Pag-Asa, a Philippine-occupied island, using DX0 callsigns and it is feasible that there will, in the future, be Spratly Island operations which count for DXCC credit from a Vietnamese-occupied island with an XV or 3W callsign.

Similarly, if you wish to operate from the Western Sahara (the Saharan Arab Democratic Republic), you will need to obtain a licence issued by the occupying force there, which is the Polisario Front. These are the S0 callsigns which, interestingly, is a prefix *not* assigned by the ITU. An operation from the Moroccan-occupied part of the Western Sahara, using a CN callsign, will not count for Western Sahara for DXCC purposes.

In the case of Northern Cyprus, it is perfectly possible to obtain a licence from the occupying Turkish authority, who will grant you a 1B callsign, again a prefix not assigned by the ITU. However, in this case your operation will not count for DXCC - in fact it will not count for anything and you will be considered a 'pirate' by most operators.

Chapter 4 of this book deals in detail with the process of obtaining a licence country by country.

REFERENCES

European Radiocommunications Office (CEPT Recommendations T/R 61-01 and T/R 61-02):
www.erodocdb.dk
IARU Region 1 (Operating abroad advice):
www.iaru-r1.org
IARP legislation: www.citel.oas.org/iarp.asp
USA IARP application form: www.arrl.org/ FandES/field/regulations/io/#IARP

Licensing Information Directory 4

In the following list of countries, 'CEPT', 'HAREC', and / or 'IARP' appear after the names of some countries. Where they do so, the licensing procedure is eased considerably.

In the case of 'CEPT', if you hold a licence issued in any of the countries that have accepted CEPT Recommendation T/R 61-01, you do not need to apply for an operating licence: you may operate using your own callsign preceded by the prefix of the country concerned (see the section on the CEPT Licence in Chapter 3 for more detailed information).

In the case of 'HAREC', if you are also from a HAREC-issuing country, you should apply for a HAREC (*Harmonised Amateur Radio Examination Certificate*) from your home country before you leave. With this you will be able to apply for a full (not just reciprocal) amateur radio licence in the overseas country. Some countries require you to have a residence permit or be resident in the country for a certain period of time before they will issue a licence on the basis of a HAREC.

In the case of 'IARP', if you are also from a CITEL IARP-issuing country, you should apply for the IARP (*International Amateur Radio Permit*) from your home country before you leave. With the IARP you will be able to operate in the overseas country using your own callsign preceded by the prefix of the country concerned.

LICENCE DOCUMENTS REQUIRED

If, however, you are visiting one of the majority of countries that is not CEPT, HAREC or IARP, you will need to apply for a so-called reciprocal licence. The term 'reciprocal' is a bit of a misnomer, as many countries will issue a local licence or permit to the holder of an overseas licence without there being any official reciprocal licensing agreement in place. These should therefore really be called 'unilateral' licences, but I doubt the term will ever catch on! The term *'visitor's licence'* is perhaps more appropriate and is the one generally used in this book.

If you do need to apply for a visitor's licence you will need, at the very least, a copy of your home licence, copy of your passport 'personal de-

tails' page and copy of the country's entry visa (if a visa is required). In many countries *certified* copies are required, i.e. the photocopies must be certified as true copies by a Commissioner of Oaths, Justice of the Peace, magistrate, solicitor or other lawyer. Where certified copies are *known* to be required we state this, with the word *'certified'* in italics. If you are applying for your licence in person from within the country it is far safer to take the *original* documents along with you *as well as* photocopies, in case uncertified photocopies are not accepted.

In the country listings on the following pages, we assume that you will be supplying the licensing authority with the above documents as

All official radio communication callsign prefixes are allocated by the International Telecommunication Union. This is the ITU HQ building in Geneva.

a minimum. If it is known that additional documents are required, these are mentioned in the individual countries' listings.

Procedures change, officials retire (or are sacked), a change of government brings in a completely new team, licensing authorities are privatised or moved from one government department to another: in compiling a book of this nature, there will undoubtedly be cases where we are overtaken by events and the details given are out of date. However, as far as is possible, the details provided were believed to be correct as of early to mid 2008, when this book was being researched.

NATIONAL AMATEUR RADIO SOCIETIES

In many countries, and especially those with only small numbers of radio amateurs, the local amateurs will know the licensing officers personally and therefore may be able to assist people trying to obtain a visitor's licence. If you already know an amateur in the country concerned it will always pay dividends to ask them about licensing first. If you don't know any amateurs in the country you are intending to visit, often the national amateur radio society may be able to help. However, in all except the largest countries, with high numbers of amateurs, these people will be unpaid volunteers. We have therefore generally *not* included contact details for the national amateur radio societies of the world in the following section of the book. Exceptions are made when it is known that visitor's licensing is carried out through the national amateur radio society (as in Azerbaijan, Indonesia and Jordan, to name but three), or if no current up-to-date information is otherwise available.

The International Amateur Radio Union (IARU) keeps an up-to-date list of its member societies, along with contact information for their officers, on its website at www.iaru.org/iaru-soc.html Once again, you are reminded that the great majority of these individuals are unpaid volunteers, therefore only contact them with licensing queries if you are serious about operating from their country and if you really need their assistance.

THE COUNTRIES AND TERRITORIES OF THE WORLD: A - Z

AFGHANISTAN

Political developments in Afghanistan over the last five to 10 years have meant that amateur radio licensing has not had a high priority. Nevertheless, amateur radio licensing has been carried out on an *ad hoc* basis for the last few years and licences are being issued to Westerners resident in the country, mainly those connected with the military presence there or working for United Nations agencies.

In 2004 the country's Ministry of Communications issued a Five-Year Development Plan for the [Afghan] years 1384 - 1389 (2005 - 2009) in which, in section 6, Priorities for 1385-6 (2006-7), under sub-section 6.1, Regulation, is the intention to "Conduct a Public Consultation and adopt a Rule on Amateur Radio", so it is to be hoped that amateur radio licensing will be placed on a firmer footing in the not too distant future.

In January 2005 the ARRL website reported that the Head of Spectrum Management Department at the then Ministry of Communications (now renamed MCIT) was Nader Shah Arian, who was very helpful. His contact details were: Spectrum Management Department office tel: +93 20 210 1130; e-mail: arian@mcit.gov.af

In 2006, the Afghan Telecommunication Law established the Afghanistan Telecom Regulatory Authority (ATRA) within the framework of the new Ministry of Communications and Information Technology (MCIT, website: www.mcit.gov.af), 6th Floor, ICT Directorate, Mohammad Jan Khan Watt, Kabul; tel: +93 20 210 1107; fax: +93 20 210 1708. ATRA now has the responsibility of regulating the telecommunications sector. Its website was very much 'under construction' at the time of going to press: there are clearly plans for a 'Radio Amateur Regulation' page under 'Spectrum Management' but, along with many others, this page did not exist as this book was being compiled.

Licensing authority: Afghanistan Telecom Regulatory Authority (ATRA), 10th Floor, ATRA Office, MOC Headquarters Tower, Mohammad Jan Khan Watt, Kabul, Afghanistan;
website: www.atra.gov.af
Tel: +93 20 210 1179.
E-mail: contact@moc.gov.af

ALBANIA

The Albanian licensing authority is the Telecommunication Regulatory Entity (Enti Rregullator i Telekomunikacioneve, ERT, in Albanian), which has a website with pages in English. As far back as 1997 Mr Hydajet Kopani was the Director General of the Albanian PTT and responsible for issuing amateur radio licences. Now, in 2008, the ITU lists Mr Kopani as the Chairman of the Board of Directors of ERT.

The Albanian Amateur Radio Association, AARA (PO Box 1501, Tirana) may be able to help visitors to Albania to receive a licence. Contact the IARU Liaison Officer, Marenglen 'Geni' Mema, ZA1B, tel / fax: +355 42 64738; e-mail: genimema@atnet.com.al

Licensing authority: Enti Rregullator i Telekomunikacioneve, ERT, Reshit Çollaku Str No 43, Tirana, Mail Box: 253/1, Albania;

The MPT building in Algiers.

website: www.ert.gov.al/ert_eng/Index.html
Tel: +355 42 59571 / +355 42 32131;
fax: +355 42 59106 / +355 42 32954.
E-mail: info@ert.gov.al

ALGERIA

It is difficult, but not impossible, to obtain a 7X0 visitor's licence in Algeria. A licence is more likely to be issued to long-term visitors, e.g. those with work permits, than to those on short visits such as a holiday or for a DXpedition. If you wish to try, you are advised to go through the IARU national society, Amateurs Radio Algeriens (ARA), PO Box 1, 16000 Alger Gare, Algeria; website: www.chez.com/7x2ara; tel / fax: +213 21 725 013; e-mail: radioara@hotmail.com
Licensing authority: Ministere de la Poste et des Technologies de l'Information et de la Communication, 4 Bd Krim Belkacem, Alger 16027, Algeria; **website:** www.mptic.dz
Tel: +213 21 711220.
E-mail: contact@mptic.dz

ANDORRA

The Andorran government only issues amateur radio licences to residents of Andorra. It is *not* normally possible for non-residents of Andorra to obtain a licence.

Andorra has not accepted CEPT Recommendation T/R 61-01 and the use of your home callsign with the C3 prefix is therefore not permitted.

According to a 2003 letter from Joan Sauri, C31US, the President of Unió de Radioaficionats Andorrans (URA, the Andorran national amateur radio society), the *only* exception is catered for under Article 7 of the regulations governing the use of amateur radio in Andorra. In this article, a special exception allows the issuing of a short-term licence for special projects involving the activation of bands and modes not normally activated by resident amateurs. Such projects are carried out in cooperation with members of URA and with the involvement of local amateurs in order to expand their knowledge and skills. The regu-lations do not permit the issuance of licences to visitors for HF and VHF expeditions, mobile operations, contest operations or other temporary activities. An example of the kind of project that was approved is an EME (moonbounce) and MS (meteor scatter) expedition.
Licensing authority: Servei de Telecomunicacions d'Andorra (STA), Carrer Mossèn Lluís Pujol 8-14, Santa Coloma, AD500 Andorra la Vella, Andorra.
Tel: +376 875 105; **fax:** +376 725 003.
E-mail: info@sta.ad

ANGOLA

Angola is one country where it is quite possible to obtain a licence if applying in person, but almost impossible in advance.

In addition to the usual documentation (see 'Licence Documents Required' above) you need an application letter *written in Portuguese*. In the letter you should state the period for which the licence is required, the location of the station (note that portable operation is not permitted and only one fixed location is allowed), the make, model and serial number of the transmitter or transceiver, and the modes and frequencies required (IARU Region 1 bands).

The power limit is negotiable, but normally 100 to 150W. The licence fee is also negotiable, approximately USD $150 for a six-month period.

To apply, go in person with all the documents and the licence fee to the Angola Institute of Communications (INACOM) office on the seventh floor of the Mutamba Building on Avenida de Portugal (formerly known as Rua Frederich Engels) in Luanda.
Licensing authority: Angola Institute of Communications (INACOM).
Postal address: INACOM, Caixa Postal 1459, Luanda, Angola.
Street address: INACOM, 7th floor, Mutamba Building, Avenida de Portugal 92, Luanda.
Tel: +244 222 338 352; **fax:** +244 222 339 356.
E-mail: info@inacom.og.ao

ANGUILLA

It is easy to get an amateur radio licence in Anguilla. Contact the licensing officer, Mr Wycliffe Richardson, and make an appointment to meet him in person in The Valley, Anguilla's tiny capital, during normal office hours.

The licence fee is XCD $100 (East Caribbean dollars), or USD $38, payable in cash.
Licensing officer: Mr Wycliffe Richardson, PO Box 296, The Valley, Anguilla, BWI.
Tel: +1 264 497 3269; **mob:** +1 264 235 7269.
E-mail: judge696@hotmail.com *or* djlrcoms@yahoo.com

ANTIGUA & BARBUDA

A visitor's amateur radio licence is easy to obtain on Antigua but it is best to apply in person. Visitor's licences are issued for one year. They cost XCD $75 (East Caribbean dollars) or USD $30, and the call is held for three years then can be re-issued unless you renew it. You can choose from any available V26 or V25 callsigns. You can also obtain a three-year licence for $175 EC, or USD $80.

The Assistant Telecommunications Officer for Antigua is Mr William Henry. His office is on the fourth floor of the State Insurance Building on the corner of Thames Street and Long Street (formerly the Cable and Wireless building) in St John.

To get your licence, go to the Telecommunications Office with your passport and a copy of your home licence. Mr Henry will then give you an invoice to take to the Inland Revenue office to pay for the licence. The Inland Revenue Office is on Newgate Street (close to the police station in a yellow building with a Shoul's Department Store sign on the top). Go to the second floor to the window with the sign for telecommunications licences, pay for your licence and get the receipt. Then return to the Telecommunications Office with the receipt and Mr Henry will print out your licence for you.

Licensing authority: Ministry of Information Broadcasting and Telecommunications, Telecommunication Division, 4th Floor State Insurance Building, Corner of Thames and Long Street, St John's, Antigua and Barbuda;
website: www.antigua.gov.ag
Tel: +1 268 562 1868; **fax:** +1 268 462 3225.
E-mail: william.henry@antigua.gov.ag

ARGENTINA (IARP)

If you are not from an IARP country, you will need to apply for an Argentinean visitor's licence. An official application form from the Argentine Comision Nacional de Comunicaciones (CNC) in Spanish and English can be found on the GACW (Grupo Argentino de CW) website at http://gacw.no-ip.org/formcnc.zip

Complete the form and send it with *certified* copies of your licence, passport and visa (if required) to an official Argentine radio club. A full list of the 188 official radio clubs can be found on the CNC website at www.cnc.gov.ar/espectro/radioaficionados/RadioClubes.asp Your licence must be written in Spanish, Portuguese or English: if not, an *official* translation must be attached. All copies must be certified by a public authority.

You should apply 30 - 40 days before your licence is required. You will be granted the equivalent class of licence to your home licence, but you must observe local regulations, e.g. the 30m band

in Argentina is at 10110 - 10131.5kHz only. If you are granted a 'General' or 'Superior' class of licence you may operate from an existing contest or expedition station using a special callsign, but not if you are issued a 'Novice' or 'Intermediate' licence. The callsign issued will be LU/own call: it is not possible for visitors to obtain an LU, LR, LW, L2 or AY callsign. There is no charge for an Argentine visitor's licence. (Information courtesy of GACW)

Licensing authority: Comision Nacional de Comunicaciones (CNC), Seccion Radioaficionados, Peru 103, Piso 12, C1087AAC - Buenos Aires, Argentina; **website:** www.cnc.gov.ar/espectro/radioaficionados
Tel: +5411 4347 9641 / 9642;
fax: +5411 4347 9711.
E-mail: radioaficionados@cnc.gov.ar

ARMENIA

To obtain a visitor's licence in Armenia you should apply through the country's national amateur radio society, the Federation of Radiosport of the Republic of Armenia (FRRA). Documents required are a completed application form and a copy of the licence issued by the country of which you are a citizen. A blank application form pdf can be found on the ARRL website at www.arrl.org/FandES/field/regulations/io/armenia-application.pdf

The documents should be submitted at least two months before the licence is required. It is understood that the licence fee is approximately USD $250 for a two month licence. You need a separate permit to bring equipment into the country. For further details contact: George Badalian, EK6GB, President of the Federation of Radiosport of the Republic of Armenia (FRRA), 87 Arshakuniats Ave, 375007 Yerevan, Armenia; tel: +374 1026 7829; tel / fax: +374 156 5616; mob: +374 9145 8354; e-mail: ek6gb@mfa.am

Licensing authority: Ministry of Transport and Communications, 28 Nalbandyan Street, 375010 Yerevan, Armenia.
Tel: +374 10 563 391; **fax:** +374 10 560 528.
E-mail: ministry@mtc.am

ARUBA

Although part of the Kingdom of the Netherlands, Aruba is *not* a member of CEPT (nor CITEL's IARP), so it is necessary to apply for a visitor's licence. Licensing information for Aruba can be found on the website of the Aruba Amateur Radio Club (AARC) at www.qsl.net/aarc/w_p4_license.htm The following is a summary. Download the application form on the AARC website and send it by fax to the Direktie Telecommunicatie Zaken (DTZ). You need to apply at least two months before the licence is required.

The licence fee is AWG 108 (Aruban guilders, also known as florins, approx GBP £31). The fee for a P40 special contest call is an additional AWG 50 (approx £14).

If you are taking your own equipment and operating from an hotel, you can also expect a station inspection. The charge for inspections is AWG 100 (approx £28). If you are using an existing station this charge is waived if the station owner has had a recent inspection.

All licence fees must be paid at a local bank as the DTZ does not accept payments over the counter. You will receive an invoice by fax or by post, and should pay this at a local bank. Then take the bank receipt to the DTZ office (office hours: 8.00am - 5.00pm, Monday to Friday) as proof of payment and the licence will be issued.

Licensing authority: Direktie Telecommunicatie Zaken (DTZ), Caya Betico Croes 149, Oranjestad, Aruba.

Tel: +297 582 6069; **fax:** +297 582 5307.

ASCENSION ISLAND

ZD8 licences are issued on the spot and you have a choice of any callsign that is still available. Apply to the Administration Department of the Ascension Island Government.

Licensing authority: Ascension Island Government, Georgetown, Ascension Island ASCN 1ZZ; **website:** www.ascension-island.gov.ac

Tel: +247 7000 ext 100; **fax:** +247 6152.

E-mail: aigenquiries@ascension.gov.ac

AUSTRALIA (CEPT, HAREC)

Amateur radio licensing in Australia is carried out by the Australian Communications and Media Authority (ACMA). Licensing for overseas visitors was liberalised considerably with effect from 14 February 2008, when a new so-called 'Class Licence' came into effect. As a result of this, the ERO added Australia to the list of CEPT Licence countries in May 2008.

However, even for those without a CEPT Licence, the Australian Class Licence makes it very easy indeed for almost any radio amateur to operate from Australia. It authorises operation for up to 90 days after each entry to Australia but if the operation of the amateur station starts or finishes more than 90 days after the overseas amateur enters Australia, that operation must be authorised by an 'Apparatus Licence' (normal visitor's licence, see below). For example, if an overseas amateur visits Australia for a period of 91 days he or she may operate an amateur station for the first 90 days under the Class Licence and not operate the amateur station on the 91st day. Alternatively, he or she may apply for an Apparatus Licence to cover the entire 91 days.

The Class Licence applies in Australian territories, e.g. the Cocos (Keeling) Islands, Norfolk Island etc, in the same way it applies to mainland Australia.

The Class Licence provides for five different levels of operation. Each of these levels corresponds to the qualification or licence held by the overseas amateur. In practice, this means that holders of some overseas novice-type licences may operate, subject to certain conditions, in Australia.

The callsign to be used by amateurs utilising the Class Licence is their own call/VK, e.g. M0QQQ/VK (although it is unclear whether or not the call district number should be appended to the callsign, e.g. /VK2 for New South Wales or /VK9 for Australian territories).

Details of the Class Licence can be found on the ACMA website under "Amateurs visiting Australia" at www.acma.gov.au/WEB/STANDARD/pc=PC_1311 and the full text of the "Radiocommunications (Overseas Amateurs Visiting Australia) Class Licence 2008", a 37-page pdf document, can be also be downloaded here.

If you are visiting Australia for more than 90 days, you will need to apply for an Australian visitor's licence in the normal way. Australia has also accepted CEPT Recommendation T/R 61-02 so if you are from a country that issues HARECs you may obtain a full Australian licence (not merely a reciprocal) on the basis of your HAREC.

If you are not from a HAREC country, Australia has reciprocal licensing agreements with 17 countries (Canada, Denmark, France "including New Caledonia", Germany, Greece, India, Israel, Japan, Malaysia, New Zealand, Papua New Guinea, Poland, Solomon Islands, Spain, Switzerland, United Kingdom and USA) and around 20 more where, although there are no official reciprocal arrangements in place, the countries have a class of certificate or licence accepted as equivalent to an Australian qualification.

In most cases, Australian amateur licences are issued for the period of an overseas amateur's stay in Australia, as noted on the person's visa. The following documents need to be provided when applying for a licence (with English translations to be supplied where applicable):

• a copy, certified by a public notary, of the applicant's current licence or certificate of qualifications;

• a certified copy of the applicant's passport;

• proof of the duration of the visit, such as a visa, or if issued with an Electronic Travel Authority, a certified true copy of a "travel ticket to Australia";

• a completed licence application form entitled Application for Apparatus Licence(s) (R057), which is available on the ACMA website;

• the current licence fee (in Australian dollars).

Applications should be lodged at least three months prior to arrival in Australia. This will allow enough time for the licence to be issued and forwarded prior to arrival.

Licence fees were amended in July 2007. The fee is made up of an issue or renewal charge of AUD $23 plus a tax amount of $38, making a total of $61 (approx GBP £27.50) for up to one year. However, visitors from some countries are now issued with an Electronic Travel Authority (ETA) instead of a traditional visa. For visitors travelling under an ETA, an amateur licence will only be issued for a maximum of three months from the date of entry into Australia.

Further information for visiting radio amateurs can be found on the ACMA website at www.acma.gov.au/WEB/STANDARD/pc=PC_1311

Licensing authority: National Licensing and Allocations Branch, Australian Communications and Media Authority, Canberra Central Office; **website:** www.acma.gov.au
Postal address: PO Box 78, Belconnen ACT 2616, Australia.
Street address: Purple Building, Benjamin Offices, Chan Street, Belconnen, ACT 2617.
Tel: +61 2 6219 5555; **fax:** +61 2 6219 5200.
E-mail: aas@acma.gov.au

AUSTRIA (CEPT, HAREC)

Austria has accepted both CEPT Recommendation T/R 6-01 and T/R 61-02 and therefore amateur radio operation by visitors is not a problem for many nationalities. If you do not have a CEPT licence or a HAREC, you will need to apply for an Austrian visitor's licence. An application form in English can be downloaded from the website of the Austrian national amateur radio society, OeVSV, at www.oevsv.at/opencms/funkbetrieb/gastlizenz.html When completed it should be sent with a copy of your home country's licence to the licensing office. The OeVSV offers assistance if you e-mail them at info@oevsv.at
Licensing authority: Fernmeldebüro für Wien, Niederösterreich und Burgenland, Höchstädtplatz 3, A-1200 Wien (Vienna), Austria.
Tel: +43 1 3318 1112; **fax:** +43 1 334 2761.

AZERBAIJAN

Applications for a visitor's licence in Azerbaijan must be made through the national amateur radio society, the Federation of RadioSport of Azerbaijan (FRS). The FRS is entitled to test applicants' knowledge of radio theory and Morse code, but usually a full licence from an overseas country is accepted as sufficient proof of knowledge. However, novice or similar class licensees can expect to be tested.

The following documents are required: two copies of the completed application form, one copy of your home licence, one copy of your passport, two copies of your CV (resume), two photographs 4 x 6cm in size, and (if you are working in Azerbaijan) a certificate confirming that from your employer. The application form may be obtained from the President of the FRS, Natig Kasimov, 4J5T.

The callsign issued to visitors has either 4J or 4K as the prefix, followed by the digit 0 or 1. 4J/4K0 and 1 callsigns are only issued to visitors, Azerbaijani DXpeditions and stations of the Azerbaijan Amateur Radio Emergency Network (AzAREN). There are three classes of licence: Category 1 (all bands, all modes, power 200 watts, one or two letter suffixes), Category 2 (all bands except 10, 18, 24MHz, all modes except no phone on 14MHz, power 50 watts, two letter suffixes) and Category 3 (CW on all bands except 7, 10, 14, 18 and 24MHz, phone only on the 1.8, 3.6 and 28MHz, power 10 watts, three-letter suffixes).

The licence fee is dependent on the class of licence issued, but is not more than AZN 10 Azerbaijan New Manats (approx GBP £6) for up to one year. If more than one year is required, a permanent licence has to be obtained with a regular (not special) callsign.

The application should be sent to Natig Kasimov, 4J5T, the President of the FRS, tel: +994 12 493 6654; mob: +994 50 210 0023; e-mail: teknatig@azercell.com or natigkasimov@gmail.com
Licensing authority: State Department for Radio Frequency Usage, Ministry of Communications and Information Technologies (MCIT), 33 Azerbaijan Avenue, PO AZ1000, Baku, Azerbaijan;
website: www.mincom.gov.az/en/main.html
Tel: +994 12 498 5838;
fax: +994 12 498 7912 / 8019.
E-mail: mincom@mincom.gov.az

BAHAMAS

Visitor's licences are granted to individuals from countries with which the Bahamas has reciprocal agreements. You should submit a written application indicating the location of the amateur radio station and the period of operation and enclose a copy of your passport or birth certificate showing place and date of birth, a copy of your home licence, and an international money order for USD $25.00, made out to the 'Public Utilities Commission'. US licensees should note that the American Technician licence class is not accepted in the Bahamas.

The application should be sent to the Public Utilities Commission at least a month before the licence is required: processing of reciprocal amateur radio licences normally takes approximately two weeks following receipt of all the required information. Any changes to this procedure will be posted on the PUC website.

Licensing authority: Radio Licensing Department, Public Utilities Commission, Agape House, 4th Terrace East Collins Avenue, PO Box N 4860, Nassau, Bahamas; **website:** www.pucbahamas.gov.bs/radio_amateur.php
Tel: +1 242 322 4437; **fax:** +1 242 323-7288.
E-mail: info@PUCBahamas.gov.bs

BAHRAIN

It is not possible for short-term visitors to obtain an amateur radio licence in Bahrain. Foreign radio amateurs holding a Bahrain residence permit, however, can apply for a licence. All applications for licences must be made through the Amateur Radio Association of Bahrain (ARAB), the IARU member society. The applicant must be a fully paid up member of ARAB. Application forms are available from the Association.

The cost is made up as follows: annual membership of ARAB BHD 12,000 dinars; application fee with the Telecommunications Regulatory Authority (one-off payment) 10,000 dinars; annual licence fee 12,000 dinars, making a total of 34,000 dinars (approx GBP £46).

The application should be sent to Amateur Radio Association of Bahrain, PO Box 22381, Muharraq, Bahrain.

Licensing authority: Telecommunications Regulatory Authority (TRA); **website:** www.tra.org.bh
Postal address: PO Box 10353, Manama, Kingdom of Bahrain.
Street address: 7th Floor, Taib Tower (opposite Embassy of the State of Kuwait), Diplomatic Area, Manama.
Tel: +973 1752 0000; **fax:** +973 1753 2125.
E-mail: contact@tra.org.bh

BANGLADESH

It has not been possible to sit for a licence exam in Bangladesh since 2004. However, thanks to the efforts of the national society, the Bangladesh Amateur Radio League (BARL), exams were re-introduced in August 2008. What effect this will have on visitors' licensing is at present unclear. Before August 2008, licensing for visitors was possible, though difficult. In theory the Bangladesh Telecommunication Regulatory Commission (BTRC) would issue a seven-day licence to radio amateurs from any country, although the application could take up to six months to be processed. It was sometimes possible to obtain a more permanent licence after the seven-day licence expired.

The BARL can provide licence application forms. You also need to provide a covering letter requesting the licence and advising the BTRC of the dates the licence is required, the dates you will be in Bangladesh (if different) and the name and address of any local contacts (e.g. BARL members), as well as copies of your home licence, passport and Bangladesh visa.

For further details, contact BARL, c/o Anwar Islam, S21L, Basati Castle, Flt #C/9, H#8/A/Kha, Rd#14, DRA, Dhaka 1209, Bangladesh (tel: +880 2 815 0533 or +880 17 1185 6629; e-mail: s21l@barl.org) well before your intended visit. The BARL has a website at www.barl.org

Licensing authority: Bangladesh Telecommunication Regulatory Commission (BTRC), Setu Bhaban (4th & 5th Floor), New Airport Road, Banani Dhaka-1212, Bangladesh;
website: www.btrc.gov.bd
Tel: +880 2 989 3917- 21;
fax: +880 2 989 0029 / 0040.
E-mail: btrc@btrc.gov.bd

BARBADOS

In Barbados, amateur radio licensing is handled by the Telecoms Unit, a department of the Ministry of Economic Affairs & Development. In February 2005 the Minister responsible for Telecommunications announced the commencement of Phase III of the Telecommunications Liberalisation timetable and that new licences - including a "licence to keep, install, erect and use an amateur radio transmitter" were available.

The Telecoms Unit has an application form for an amateur licence (Form TU 039) on its website, but this is intended for citizens of Barbados. Visitor's licences are available 'over the counter' upon application with the usual documentation. The licence fee is BBD $30 Barbados dollars (approx GBP £7.70).

Licensing authority: The Chief Telecommunications Officer, Telecoms Unit, 3rd Floor East, The Warrens Office Complex, Warrens, St Michael, Barbados; **website**: www.telecoms.gov.bb
Tel: +1 246 310 2251 / 2266;
fax: +1 246 426 0960.
E-mail: Reginald_Bourne@Barbadosbusiness.gov.bb *or* Esther_Roach@Barbadosbusiness.gov.bb

BELARUS

Visitors to Belarus should apply more than two months before the licence is required. An application form can be found on the website of Vladimir 'Walt' Sidorov, EU1SA, the President of the Belarussian Federation of Radioamateurs and Radiosportsmen (BFRR, the IARU member society), at www.qsl.net/eu1sa/travel.htm The licence fee is USD $12 and the callsign issued is EW/own call.

Send the application form, a copy of your home

licence and proof of payment by post or fax to: UP BelGIE, 22 Engels St, Minsk 220030, Belarus; fax: +375 17 222 4783. Walt advises, *"If you have friends in Belarus, you can ask them to make the payment locally and in the local currency. By doing that you can save on banking charges."* Walt also offers to help with questions related to amateur radio in Belarus. His contact details are: Vladimir V Sidorov, 220050 Minsk, PO Box 474, Belarus; tel / fax: +375 17 289 3045 (office); mob +375 296 474 474; e-mail: eu1sa@belsonet.net

Licensing authority: Ministry of Communications and Information, 10 Independence Avenue. 220050 Minsk, Belarus.
Tel: +375 17 227 3861; **fax:** +375 17 226 0848.
E-mail: mpt@mpt.gov.by

BELAU – SEE PALAU

BELGIUM (CEPT, HAREC)

If you are from a country with a CEPT licence you may operate for up to three months in Belgium using the prefix ON/ before your own callsign. If you are staying longer than three months and if you are from a country that issues HARECs, you may apply for a full Belgian callsign using your HAREC.

If, however, you do not have a CEPT licence or a HAREC, you must apply for a visitor's licence. Send a letter of application with the usual documentation to the Belgian Institute for Postal service and Telecommunications (BIPT / IBPT). The BIPT website has pages about amateur radio licensing in Dutch, French and English. The English page is at: www.bipt.be/en/274/ShowContent/1107/Radio_amateurs/Radio_amateurs.aspx

Licensing authority: BIPT, Dienst Vergunningen, Sterrenkundelaan 14 bus 21, B-1210 Brussels, Belgium; **website:** www.bipt.be
Tel: +32 22 26 88 25; **fax:** +32 22 26 88 03.
E-mail: radiovergunningen@bipt.be

BELIZE

Amateur radio licensing in Belize is carried out by the Public Utilities Commission (PUC). You can apply by post (at least one month before the licence is required), or in person at the PUC office in Belize City.

Apply with copies of your home licence and passport. The licence fee is USD $20, to be sent by international money order if applying by post or in cash if applying in person.

The Telecommunications Office can also issue an import permit for your transmitting equipment: provide a list with the make, model and serial numbers of all equipment you intend to take into the country and ask for an import permit to be issued at the same time as your licence.

The Manager of Telecommunications (2008) is Mr Kingsley Smith.

Licensing authority: Telecommunications Office, Public Utilities Commission, 41 Gabourel Lane, PO Box 300, Belize City, Belize.
Tel: +501 223 4938; **fax:** +501 223 6818.
E-mail: puctelcom@puc.bz

BENIN

In 2004, when OPT (Office des Postes et Telecommunications du Benin) was the licensing authority, amateur radio licences were issued in about three months. Five copies of an application form and your home licence were required.

The new regulatory authority is HAAC. Their website was largely 'under construction' as this book was being compiled.

Licensing authority: Haute Autorité de l'Audiovisuel et de la Communication (HAAC), Boulevard de la Marina, 01 BP 3567, Cotonou, Benin;
website: www.haacbenin.org
Tel: +229 21 311 743; **fax:** +229 21 311 742.
E-mail: infohaac@haacbenin.org

BERMUDA

The Bermuda Department of Telecommunications offers a visitor's licence at no cost to the visiting radio amateur. Just send a copy of your licence, along with the dates of your visit and the address at which you will be staying, to them by fax or walk into their office in Hamilton and present your information to them in person. The process usually only takes about 15 minutes.

If you're from the USA note that only General Class, and above, licensees may operate on Bermuda. It is also important to note that the power restriction of 100 watts applies to *all* licensees.

Further information for visiting radio amateurs, as well as the complete Bermuda amateur radio regulations, are available on the Radio Society of Bermuda website at www.bermudashorts.bm/rsb

Licensing authority: Department of Telecommunications; **website:** www.mtec.bm
Postal address: Department of Telecommunications, PO Box HM 101, Hamilton HM AX, Bermuda.
Street address: Department of Telecommunications, F B Perry Building, 2nd Floor, 40 Church Street, Hamilton HM12, Bermuda.
Tel: +1441 292 4595; **fax:** +1441 295 1462.
E-mail: gtelecom@gov.bm

BHUTAN

Unless you have a work permit and will be resident in Bhutan, the only way to get an amateur radio licence is to book a package holiday in the country and pay the obligatory daily 'Minimum Tourist Tariff' of approximately USD $200 per person per night (fixed by the government's Department of Tourism).

Licences are issued by the Bhutan InfoComm and Media Authority (BICMA). Visitors must apply for a licence and callsign, either through their appointed tour operator in Bhutan or directly to BICMA. BICMA will process the application and grant a visitor's permit to eligible amateurs for a period up to the duration of their visa, but not exceeding three months. Foreign operators requesting permission should submit a photocopy of their home licence along with the application form.

Proteus Tours, which operates the Bhutan Ham Centre, has an amateur radio licence application form on its website at www.proteustours.com.bt/ham/index.html and claims "callsign and radio licence processed in *one day* - guaranteed!" (see the 'Rental Stations' section of this book for further details of the Bhutan Ham Centre.)

If you wish to 'do-it-yourself', Form 1, an application form for 'Radiocommunication and Spectrum Licence / Permit' (which includes amateur licences) can be found on the BICMA website at www.bicma.gov.bt/form/Radioform.pdf Note that the application will not be processed unless the appropriate fee is included with the application form.

The callsign issued is in the series A52xx, where xx are any two letters of your choice. The licence fee is in addition to the daily tariff (which includes accommodation, transport, food etc) and is USD $500 for a group, $100 for an individual, with a 'High Power Transmission Fee' (over 100W) of $150 per individual.

Licensing authority: Bhutan InfoComm and Media Authority (BICMA), Royal Government of Bhutan, Post Box 1072, GPO Thimpu, Bhutan; **website:** www.bicma.gov.bt
Tel: +975 2 321 506 / 507; **fax:** +975 2 326 909.
E-mail: bicma@druknet.bt

BOLIVIA

Mario Iberkleid, CP1FF, says that Radio Club La Paz can help with visitor's licences; tel: +591 2 222 2069; e-mail: radioclublapaz@yahoo.com

Mario comments, *"The laws here are a bit unstable due to people not knowing how amateur radio works, and the government has to issue the licence. I am always willing to lend a hand for some help."*
His contact details are: Mario Iberkleid, CP1FF, PO Box 764, La Paz, Bolivia; tel: +591 2 282 4040 (office); tel: +591 2 279 9523 (home); mob: +591 772 91888; fax: +591 2 279 3732 (home); e-mail: marioiberkleidz@gmail.com

Licensing authority:
Superintendencia de Telecom-unicaciones (SITTEL), Oficina Central, Calle 13 No 8260 - 8280 Calacoto, Casilla Postal 6692, La-Paz, Bolivia; **website:** www.sittel.gov.bo
Tel: +591 2 2772266;
fax: +591 2 2772299.
E-mail: info@sittel.gov.bo

BOSNIA AND HERZEGOVINA (CEPT)

Bosnia and Herzegovina accepts the CEPT Licence with the equivalence between the CEPT Licence and the highest national licence level as of September 2003, i.e. before the Morse code references were removed from T/R 61-01. In other words, a Morse code qualification is still required for the use of the HF bands.

Please note that at the time this book was being compiled, the ITU-allocated prefix for Bosnia and Herzegovina was 'gradually' being changed from T9 to E7. Visitors should check which prefix is the appropriate one to use.

If you do not have a CEPT licence, you will need to apply for a visitor's licence. Send your application to the Communications Regulatory Agency (CRA).
Licensing authority: Communications Regulatory Agency, Mehmeda Spahe 1, 71000 Sarajevo, Bosnia and Herzegovina; **website:** www.rak.ba
Tel: +387 33 250 600; **fax:** +387 33 713 080.
E-mail: info@cra.ba *or* info@rak.ba

BOTSWANA

The licensing authority is the Botswana Telecommunications Authority (BTA). There is an amateur licence application form in pdf format on the BTA website at www.bta.org.bw/application.html The licence fee is BWP 165 pula (approx GBP £13.60). As of mid-2008 the licensing officer was Mr Samuel Mpaesele.
Licensing authority: Botswana Telecommunications Authority **website:** www.bta.org.bw
Postal address: Private Bag 00495, Gaborone, Botswana.
Street address: Plot 206 and 207, Independence Avenue, Gaborone, Botswana.
Tel: +267 395 7755; **fax:** +267 395 7976.
E-mail: mpaesele@bta.org.bw

BRAZIL (IARP)

Brazil accepts the International Amateur Radio Permit but if you are not from an IARP-issuing country you may apply to the licensing authority, ANATEL, for a visitor's licence. ANATEL's website (in Portuguese) has an amateur radio section.
Licensing authority: Agência Nacional de Telecomunicações (ANATEL), SAUS Quadra 6 - Bloco H, 70070-940 Brasília, DF, Brazil;

Brazil, PY, call districts.

website: www.anatel.gov.br
Tel: +55 61 2312 2063; **fax:** +55 61 2312 2244.

BRITISH INDIAN OCEAN TERRITORY (CHAGOS ISLANDS)

Almost all amateur radio operation from the British Indian Ocean Territory (BIOT) is by US military personnel from the base on the island of Diego Garcia. Diego Garcia has very limited access and only personnel connected with the military base are allowed to land on the island.

VQ9 amateur radio licences are issued by the British Representative on Diego Garcia by showing your home licence. You may choose any VQ9 callsign if it has not already been issued.

All persons, except those connected with the military facility on Diego Garcia, require a permit to visit BIOT. Any permit issued covers access to the islands of the Territory, except Diego Garcia

PHOTO: US DEPARTMENT OF DEFENSE

The island of Diego Garcia in the British Indian Ocean Territory.

and those designated as "strict nature reserves". It *is* therefore possible to visit some of the outer islands of the Territory if you have a permit, although these islands are only accessible by sea, i.e. private yacht. Permits are issued by the British Indian Ocean Territory Administration in London and an application form can be found on the Foreign and Commonwealth Office website. Once in possession of a BIOT entry permit, it should then be possible to obtain a VQ9 amateur radio licence.

Licensing authority: British Indian Ocean Territory Administration, Overseas Territories Directorate, Foreign and Commonwealth Office, King Charles Street, London SW1A 2AH, UK;
website: www.fco.gov.uk
Tel: +44 20 7008 2890/2691;
fax: +44 20 7008 1589.
E-mail: BIOTadmin@fco.gov.uk

BRITISH VIRGIN ISLANDS

Visitor's licences can be applied for by post or in person. A licence application form and a payment by credit card authorisation form can be found on the OH2MCN website at www.qsl.net/oh2mcn/vp2v_a.pdf and www.qsl.net/oh2mcn/vp2v_auth.pdf respectively.

In addition to the application form and payment you must provide notarised (*certified*) copies of your passport (or birth certificate) and your current home licence. If you apply in sufficient time, your licence can be posted to you in advance of your visit.

The licence fee is USD $20. Apply to the Telecommunications Manager, Mr Gregory Nelson.
Licensing authority: Telecommunications Manager, Government of the British Virgin Islands, R G Hodge Plaza (Upstairs Vehicle Licensing Dept), Road Town, Tortola, British Virgin Islands.
Tel: +1 284 468 3603, ext 3092, 3093;
fax: +1 284 494 6462 / 3934.
E-mail: ganelson@bvigovernment.org, gnelson@gov.vg *or* telecomm@bvigovernment.org

BRUNEI

Amateur radio licensing in Brunei is carried out by the Authority for Info-communications Technology Industry Brunei Darussalam (AiTi). Their website has a section on amateur radio: go to "Li-

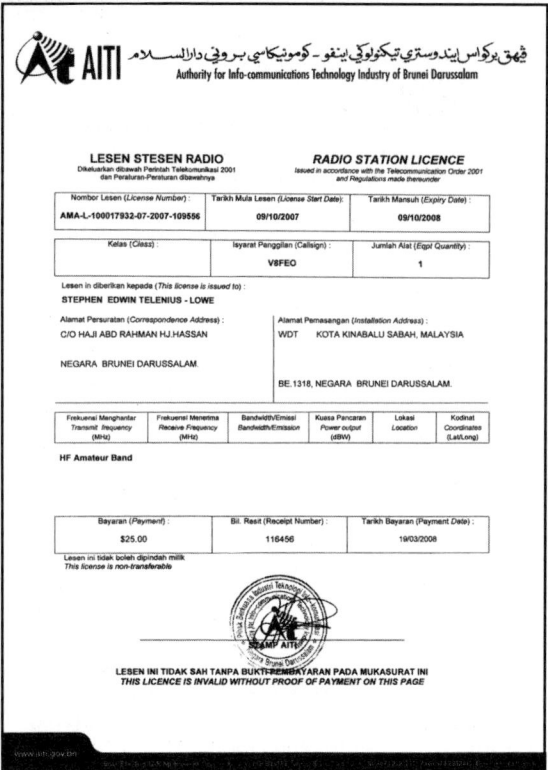

Brunei, V8, amateur radio licence.

censing" and then "Amateur Radio". A licence application form is also available for downloading.

Visitor's licences in Brunei are normally issued for a period of three months. Callsigns are in the series V8F, followed by any two letters of the applicant's choice (give three choices when applying). The licence fee is BND $25 Brunei dollars (approx GBP £9).

Hj Rahman Hassan, V85RH, is happy to help visiting foreign amateurs receive their licence. Contact him on mob: +673 873 7388.

Licensing authority: Authority for Info-communications Technology Industry Brunei Darussalam (AiTi), Block B14, Simpang 32-5, Kampong Anggerek Desa, Jalan Berakas, BB3713 Bandar Seri Begawan, Negara Brunei Darussalam, **website:** www.aiti.gov.bn
Tel: +673 2 323 232; **fax:** +673 2 382 447.
E-mail: ipgroup@aiti.gov.bn

BULGARIA (CEPT)

Bulgaria has accepted CEPT Recommendation T/R 61-01 so if you have a CEPT licence you may operate for up to three months at a time signing LZ/own call in Bulgaria. If you don't have a CEPT Licence apply to the CRC for a visitor's licence.
Licensing authority: Communications Regulation

Commission (CRC), 6 Gurko Street, 1000 Sofia, Bulgaria; **website:** www.crc.bg
Tel: +359 2 949 2335; **fax:** +359 2 987 0695
E-mail: info@crc.bg

BURKINA FASO

The licensing authority in Burkina is ARTEL, whose website (in French) has an application form for 'amateur and experimental services' available by clicking on "Documents" and scrolling to the bottom of the page.

The President of the IARU member society, the Association des Radioamateurs du Burkina Faso (ARBF), is Youssouf Kaba, XT2KY. M Kaba works for ONATEL, the Burkina Faso telecommunications provider, and may be able to help visitors to obtain a licence. Contact the Association des Radioamateurs du Burkina Faso, c/o Youssouf Kaba, ONATEL, Avenue Yennenga, 01 BP 10.000, Ouagadougou, Burkina Faso; tel: +226 300 945 / +226 363 019; fax: +226 300 930.
Licensing authority: Autorité Nationale de Régulation des Télécommunications (ARTEL); **website:** www.artel.bf
Postal address: ARTEL, Ouaga 2000, 01 BP 6437, Ouagadougou 01 Burkina Faso.
Street address: ARTEL, Avenue Dimdolobsom, Porte 43, Rue 3.48, Ouagadougou.
Tel: +226 5033 4198 *or* +226 5037 53 60/61/62; **fax:** +226 5037 5364 *or* +226 5033 5039.
Email: secretariat@artel.bf

BURMA - SEE MYANMAR

BURUNDI

For many years it has been very difficult to obtain an amateur radio licence in Burundi. Several stations were active during the 1990s, mainly operated by UN staff working in the country. However, in 1999 all operations after 1 January 1994 were disallowed by the ARRL DXCC desk when the Director General of ONATEL, the then communications authority in Burundi, informed DXCC that the licences were "forgeries".

The licensing authority is now the ARCT. A few amateur licences have been issued since 2007 and these operations have been accepted by DXCC.
Licensing authority: Agence de Régulation et de Contrôle des Télécommunications (ARCT), 360 Ave Patrice Lumumba, PO Box 6702, Bujumbura, Burundi.
Tel: +257 221 0276; **fax:** +257 242 2832.
E-mail: arct@cbinf.com

CAMBODIA

Once an extremely rare country on the bands, it is now relatively easy to take out an amateur radio licence in Cambodia. Licences are issued to indi-

Cambodia's Ministry of Posts and Telecommunications building in Phnom Penh.

viduals for one transceiver, maximum 100W, and for one location only. Portable operation is not permitted; if you wish to operate from a second location (e.g. one of Cambodia's offshore islands, which count as AS-133 for IOTA), you will need a second licence (and another callsign).

The Chief of the Frequency Licensing Office of the Ministry of Posts and Telecommunications of Cambodia (MPTC), Mr Lim Vuthy, is the only person who issues amateur radio licences. Make an appointment to visit Mr Lim and take along the following paperwork:
• a completed application form;
• a certified copy of your home licence (or take the original for inspection with the copy);
• a photocopy of your passport;
• a photocopy of your Cambodian visa;
• the technical specifications of the equipment you are planning to use.

A blank two-page application form can be found on Veikko Komppa's 'Worldwide Information on Licensing for Radio Amateurs' website at

Mr Lim Vuthy, Chief of Cambodia's Frequency Licensing Office, at his desk.

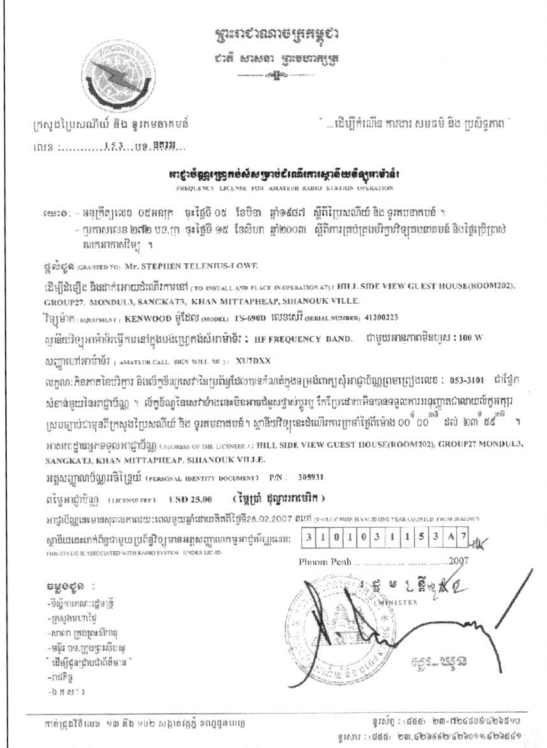

Complete with its exotic Cambodian script, this is the XU7DXX licence.

www.qsl.net/oh2mcn/xua1.jpg *and* xua2.jpg

The licence fee is USD $50 for the first year and $25 per year thereafter if you need to renew the licence. If you wish to apply for the licence by post, you should send $50 cash by registered mail to Mr Lim. The MPTC has a website which is partially in English. It has a page for the "FM & L Department" (Frequency Management and Licensing), but this was completely blank when this book was being compiled.

Licensing authority: Mr Lim Vuthy, Chief of Frequency Licensing Office, Ministry of Posts and Telecommunications (MPTC), Corner Streets 13 / 102, Sangkat Wat Phnom, Phnom Penh, Cambodia; **website:** www.mptc.gov.kh
Tel: +855 23 722 311 / 916 / +855 23 426 510 / +855 23 724 809;
mob: +855 11 876 100 (Mr Lim Vuthy);
fax: +855 23 426 011 / 012 / +855 23 428 271.
E-mail: mptc-vuthy@camnet.com.kh

CAMEROON

No specific up to date information. As with many countries, you are more likely to be successful if you apply in person rather than attempting to obtain a licence before visiting the country.

Licensing authority: Telecommunications Regulatory Board (Agence de Régulation des Télécommunications, ART), BP 6132, Yaounde, Cameroon; **website:** www.art.cm
Tel: +237 2223 0380; **fax:** +237 2223 3748.
E-mail: art@art.cm

CANADA (CEPT, IARP)

Canada recognises both the IARP and the CEPT Licence, so operating from Canada should not present a problem to most amateurs from Europe or the Americas.

If you do not have an IARP or CEPT Licence and wish to operate in Canada, you will need to apply for a visitor's licence. Visitor's licences are issued by Industry Canada to amateurs from those countries with which Canada has a reciprocal licensing agreement. The Radio Amateurs of Canada (RAC) website at www.rac.ca lists those countries as: Antigua and Barbuda, Argentina, Australia, Bahamas, Barbados, Bermuda, Bolivia, Botswana, Chile, Colombia, Costa Rica, Dominica, Dominican Republic, Ecuador, Greece, Grenada, Guatemala, Haiti, Honduras, India, Indonesia, Jamaica, Japan, Malta, Mexico, Nicaragua, Norway, Panama, Papua New Guinea, Philippines, Poland, Saint Lucia, Senegal, Slovenia, Spain, Suriname, Trinidad and Tobago, United States of America and Venezuela. (Surprisingly, the United Kingdom is not included in this list, although this is not important since Canada recognises the CEPT Licence.)

If you are applying for a visitor's licence, you should provide a photocopy of your current licence, state the level of amateur radio qualification you possess and provide an itinerary for your visit to Canada, giving the addresses where you will be staying. If applying by post, you should apply as far in advance as possible; at least a month if possible. You can also apply in person at any of the 15 Industry Canada Regional Offices, the addresses are on the Industry Canada website. You must have your licence with you if you apply in person.

The Industry Canada website has a large section about amateur radio. To navigate to it, go to www.ic.gc.ca then click on 'English', then 'By Subject' under 'Programs and Services' and finally 'Radio, Spectrum and Telecommunications'.

A map showing the Canadian call districts can be found on page 86.
Licensing authority: Industry Canada, Amateur Radio Service Centre, PO Box 9654, Postal Station T, Ottawa, Ontario K1G 6K9, Canada; **website:** www.ic.gc.ca
Tel: +1 888 780 3333 (toll free within Canada); **fax:** +1 613 991 5575.
E-mail: spectrum.amateur@ic.gc.ca

CAPE VERDE

Visitors' licensing in Cape Verde has been on again, off again. For much of the 1990s it was impossible for visitors to obtain an amateur radio licence here, although they were granted to residents. A few six-month visitor's licences were issued from 2000 onwards and, although there is no information about amateur radio on the ANAC website, it is believed that at present it is still possible to obtain a licence in Cape Verde.
Licensing authority: Agência Nacional das Comunicações (ANAC), Edifício MIT, Ponta Belém, PO Box 892, Praia, Republic of Cape Verde; **website:** www.anac.cv
Tel: +238 260 4400 / 01; **fax:** +238 261 3069.
E-mail: info@icti.gov.cv

CAYMAN ISLANDS

Licences in the Cayman Islands are issued by the Information and Communications Technology Authority (ICTA). It is possible to apply for the licence in person after arriving in the Cayman Islands but it is advisable to apply in advance so that you may also obtain an equipment import permit. This will prevent any difficulties with customs if you are importing your own equipment.

Send an e-mail to request a licence application form and credit card authorisation form. This will allow you to pay the licence fee (USD $25.00) and import permit (USD $12) by credit card. When completed, send these with copies of your passport and home licence by e-mail and your application will be processed.

The callsign issued to visitors is ZF1 followed by any two letters for Commonwealth citizens (e.g. UK amateurs), or ZF2 for non-Commonwealth visitors. If you have a preference for a specific callsign, you should indicate this on your application. (Residents on Grand Cayman also have ZF1 callsigns, whereas those on Little Cayman have the prefix ZF8 and those on Cayman Brac ZF9. Visitors' ZF1 and ZF2 callsigns may be used anywhere in the island group.) The maximum power output is 1kW.

The Licensing Analyst at ICTA is Ms Nikki S Forbes.
Licensing authority: Ms Nikki S Forbes, Licensing Analyst, Information & Communications Technology Authority (ICTA); **website:** www.icta.ky
Postal address: PO Box 2502 GT, George Town, Grand Cayman KY1-1104, Cayman Islands.
Street address: 3rd Floor, Alissta Towers, 85 North Sound Way, George Town, Grand Cayman.
Tel: +1 345 946 4282 / +1 345 945 8283 *or* +1 345 746 9613; **fax:** +1 345 945 8284.
E-mail: nikki.forbes@icta.ky *or* licensing@icta.ky

CENTRAL AFRICAN REPUBLIC

Licensing used to be carried out by the Radio

Communications Section of SOCATEL (BP 939, Avenue des Martyrs, Bangui; tel: +236 61 2264; fax: +236 61 5048), but the responsibility is now believed to have been moved to the new regulator, Agence chargée de la Régulation des Télécommunications (ART).

The following describes the application procedure when licences were issued by SOCATEL, and is probably still very similar. It is only possible to obtain a licence in the Central African Republic by applying in person. You can expect the procedure to take two to three months. Initial paperwork requires the completion of a questionnaire about the equipment, frequencies, power and antennas you plan to use. You are advised to request operation on "all amateur radio frequencies" and to specify 100 watts. You may request any TL callsign you wish when completing this questionnaire. The document is forwarded to the Presidential Security department for approval, followed by the ministries of Interior and National Safety, Post and Telecommunications, and Defence.

When the four signatures have been obtained you should provide copies of your home licence and passport. The application and licence fees are 19,000 CFA francs (approx GBP 21.50) and the licence is valid for one year.

Licensing authority: Agence chargée de la Régulation des Télécommunications (ART), Direction Générale, BP 1046, Bangui, Central African Republic.
Tel: +236 75 548292; **fax:** +236 21 610582.
E-mail: artca@intnet.cf

CHAD

Chad is another country where it was possible to get an amateur radio licence if applying in person, but not from outside the country. Troubles in Chad in early 2008 may have changed that, however, and the OTRT website was not functioning whenever checked in 2008.

Apply in person with a letter written in French requesting a licence addressed to the Director General of OTRT. If you do not speak French you are advised to take an interpreter along with you. Also take a copy of your home licence, a copy of your passport and two passport-size photos. The licence fee is 17,700 CFA (approx GBP £20).

Licensing authority: Office Tchadien de Régulation des Télécommunications (OTRT), Gestion Nationale des Fréquences, Boîte Postale 5808, N'Djamena, Chad; **website:** www.otrt.td
Tel: +235 521 250 / 513;
fax: +235 521 343 / 515.
E-mail: otrt@intnet.td

CHAGOS ISLANDS – SEE BRITISH INDIAN OCEAN TERRITORY

CHILE

To obtain a visitor's licence in Chile you can either apply direct to the licensing authority, or via the Radio Club of Chile.

An application form can be found on the OH2MCN website at www.qsl.net/oh2mcn/cea.pdf You also need a copy of your licence and your passport personal details page.

There is no licence fee, but if you ask the Radio Club of Chile for help you are requested to join the club temporarily as an overseas member. Contact the President, Radio Club de Chile, Dino Besomi, CE3PG, tel: +56 2 392 1641; mob: +56 09 879 7514; e-mail: ce3pg@mi.ce for further details.

Please note that special permission is required to operate on the 10MHz band in Chile (including Juan Fernandez and Easter Island).

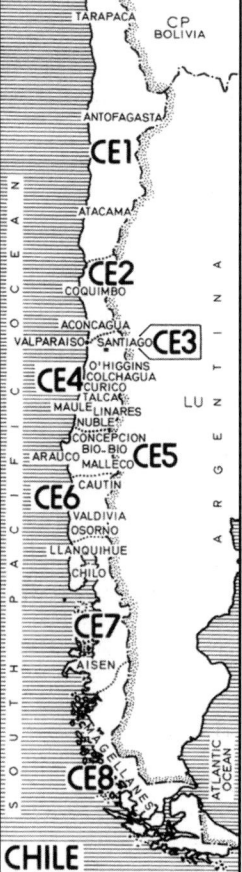

Chile, CE, call districts.

Licensing authority:
Ministerio de Trasportes y Telecomunicaciones, Departamento de Licencias de Radioaficionados, Amunátegui No 139, Correo 21, Santiago, Chile;
website: www.subtel.cl
Tel: +56 2 421 3000 / +56 2 421 3502;
fax: +56 2 421 3131 / +56 2 421 3553.
E-mail:
gabinete@subtel.cl *or* lpineiro@subtel.cl

CHINA

Although the licensing authority in the People's Republic of China is the Ministry of Information Industry (MII), visitors' amateur radio permits are issued by the Chinese Radio Sports Association (CRSA).

New regulations concerning amateur radio operation by visitors to China came into effect in 2001. Although it is not possible for short-term visitors to get their own callsign and operate their own station, since 2001 it *has* been possible to operate from Chinese individuals' private stations, as well as from club stations. In both cases, a permit must be obtained from the CRSA and the callsign to be used is "home callsign/host callsign", e.g. M0QQQ/BY1AA.

Foreigners who are living in China for more than one year may now obtain their own Chinese callsign.

In both cases, applications should be made to the CRSA, either in advance by post, or in person at the CRSA headquarters if you are in Beijing. The permit should be issued quickly. Apply with a copy of your passport, a copy of your home licence, a passport photo and the application fee of USD $5. The permit is issued for one year.

Amateur radio permits: Chinese Radio Sports Association;
website: www.crsa.org.cn/english.php
Postal address: Chinese Radio Sports Association, PO Box 6106, Beijing 100061, China.
Street address: Chinese Radio Sports Association, 14-A Tiantan Dongli Zhongqu, Beijing 100061.
Tel: +86 10 6705 0878; **fax:** +86 10 6705 0899.
E-mail: crsa@public.bta.net.cn
Licensing authority: Ministry of Information Industry, 13 West Chang'an Avenue, 100804 Beijing, PR China; **website:** www.mii.gov.cn
Tel: +86 10 6601 4670 / +86 10 6602 0051 / +86 10 6602 1330; **fax:** +86 10 6601 1370.
E-mail: info@mii.gov.cn

COLOMBIA

You are advised to apply for a licence via the Colombian Amateur Radio League (Liga Colombiana de Radioaficionados, LCRA). The callsign issued to visitors is HK, followed by a district number, then your home call, e.g. HK3/M0QQQ. The power limit is 2000 watts PEP.

To apply, write a letter in Spanish requesting a licence, and enclose a photocopy of your home licence, a photocopy of your passport clearly showing its number and expiry date, and three passport-size photographs. The application should be sent at least two months before the licence is required, preferably longer.

Send the application to: Liga Colombiana de Radioaficionados, PO Box 584, Santa Fe de Bogotá, Colombia. The President of LCRA is Ignacio Barraquer Coll, HK3CC, and he may be contacted on tel: +57 1 610 8499 or +57 1 256 3285; fax: +57 1 610 9877 or +57 1 610 4406; e-mail: ibquer@cable.net.co The LCRA has a website (in Spanish only at the time of compiling this book) at www.qsl.net/hk3lr
Licensing authority: Comisión de Regulación de Telecomunicaciones (CRT), Carrera 13 No 28-01, Piso 8, Bogotá DC, Colombia;
website: www.crt.gov.co
Tel: +57 1 327 7000; **fax:** +57 1 327 7001.
E-mail: atencioncliente@crt.gov.co

COMOROS

There is little or no amateur radio activity by residents (or visitors!) in the Comoros, but as recently as 2007 it was possible to obtain a visitor's licence. Apply in person with a copy of your passport and home licence to Comores Telecom (formerly Société Nationale des Télécommunications, SNPT) at the General Post Office in Moroni. In the past, Mr Issa Ahmed Soilihi has been the officer who issued amateur radio licences.
Licensing authority: Comores Telecom, BP 7000, Place de France, Moroni, Grande Comore, Comoros;
website: www.comorestelecom.km
Tel: +269 744 300/ +269 730 610;
fax: +269 731 079/ +269 731 016.
E-mail: comorestelecom@comorestelecom.km

CONGO (REPUBLIC – CONGO BRAZZAVILLE)

No current information about amateur radio licensing available.
Licensing authority: Direction Générale, Administration Centrale des Postes et Télécommunications (DGACPT).
Postal address: BP 2490, Brazzaville, Republic of Congo.
Street address: Avenue Paul Doumer, Brazzaville.
Tel: +242 811 693; **fax:** +242 811 695.

Colombia, HK, call districts.

CONGO (DEMOCRATIC REPUBLIC – CONGO KINSHASA)

It is possible, but both time-consuming and expensive, to obtain an amateur radio licence in the Democratic Republic of the Congo. State security clearance is required before any amateur radio licence is issued and the process can be measured in months or even years rather than weeks. The licensing procedure involves numerous meetings with officials as well as payments (both official and 'unofficial') to various ministries, the PTT and the security service. The final cost of a licence is estimated to be between USD $500 and $1000.

As a result, there are only a handful of legal licensees in DR Congo, although it is believed that other stations are operating with some form of 'unofficial' licence.

Until the situation changes, licensing is therefore unlikely to be attempted by anyone not permanently resident in the country. Those wishing to try are recommended to take advice from members of the Amateur Radio Association of DR Congo, ARAC, PO Box 2049, Hotel des Postes, Boulevard du 30 Juin, Kinshasa 1, DR Congo.

Licensing authority: Autorité de Régulation de la Poste et des Télécommunications du Congo (ARPTC), Immeuble Gécamines, Boulevard du 30 Juin, Kinshasa Gombe, Democratic Republic of Congo.
Tel: +243 139 2491 / 2493; **fax:** +243 139 2492.
E-mail: info.arptc@arptc.cd

COOK ISLANDS

Visitor's licences are issued by Telecom Cook Islands Ltd with little formality. The licence fee is NZD $20 (approx GBP £8) and the callsign series is E51 followed by three letters.

As of 2008 the licensing manager was Mr Katoa Banaba.
Licensing authority: Telecom Cook Islands Ltd, PO Box 106, Avarua, Rarotonga, Cook Islands;
website: www.telecom.co.ck
Tel: +682 29 680 ext 3283; **fax:** +682 26 174.
E-mail: sales@telecom.co.ck *or* katoabanaba@telecom.co.ck

COSTA RICA

It is necessary to obtain a permit to operate in Costa Rica. This can be obtained in San José from the Control Nacional de Radio at the Ministry of the Interior (Ministerio de Gobernacion y Policia). The permit is basically free, other than a stamp duty fee of 125 colones (less than US $1). Apply in person with a letter, preferably written in Spanish, requesting permission and stating your address or addresses in Costa Rica and the period the licence is required. You need to take copies *and originals* of your licence and passport.

The callsign issued to visitors is your home call/TI, followed by a call district number, e.g. M0QQQ/TI2 (for San José region). Full TI callsigns are only issued to residents of the country.

Note that licences for Cocos Island (TI9) are generally only issued to residents of Costa Rica. In addition to the licence, it is also necessary to obtain landing permission to visit the island.
Licensing authority: Control Nacional de Radio (CNR), Ministerio de Gobernación Policía y Seguridad Pública; **website:** www.msp.go.cr
Street address: Ministerio de Gobernación Policía y Seguridad Pública, 75m Norte del la Antigua Pulperia La Luz, Barrio Escalante (a mano derecha casa blanca), San José.
Tel: +506 2586 4000 / +506 2286 2355
E-mail: comunitaria@msp.go.cr

COTE D'IVOIRE – SEE IVORY COAST

CROATIA (CEPT, HAREC)

The use of a CEPT licence by a visitor to Croatia is restricted to a maximum period of three months at a time. However, the Croatian Amateur Radio Society, HRS, is permitted to issue a 9A8 two- or three-letter callsign to foreign nationals who apply for one. The callsign is issued by HRS under "amateur radio communications regulations, official gazette vol 198/03 and 3/04", as well as HRS's own regulations. HRS then provides a list of all issued callsigns to the Croatian Telecommunications Agency (CTA) each month.

Contact HRS at: Croatian Amateur Radio Society, Dalmatinska 12, 10000 Zagreb, Croatia. There is an e-mail message form on HRS's website at www.hamradio.hr

If you do not hold a CEPT licence or a HAREC, you could also apply direct to CTA for a reciprocal licence. An application form in Croatian, English and German can be found on OH2MCN's website at www.qsl.net/oh2mcn/9aa.htm
Licensing authority: Croatian Telecommunications Agency (CTA), Jurisiceva 13, PO Box 162, 10 002 Zagreb, Croatia; **website:** www.telekom.hr
Tel: +385 1 489 6000; **fax:** +385 1 492 0227.
E-mail: info@telekom.hr

CUBA

There are no reciprocal licensing agreements in place and, according to Cuba's Radio Regulations, foreign citizens cannot apply for an amateur radio licence. However, it *is* possible for foreign amateurs to operate from Cuba in cooperation with Cuban amateurs as a member of an organised group.

According to Cuban regulations, foreign amateurs may operate in certain "activities" organised by the Federación de Radioaficionados de Cuba

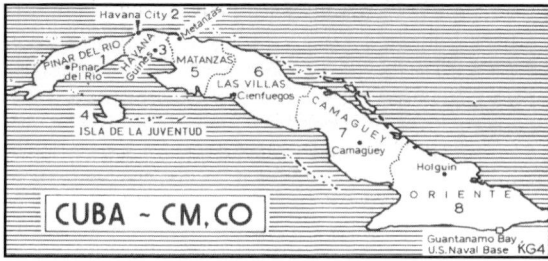

Cuba, CO, call districts.

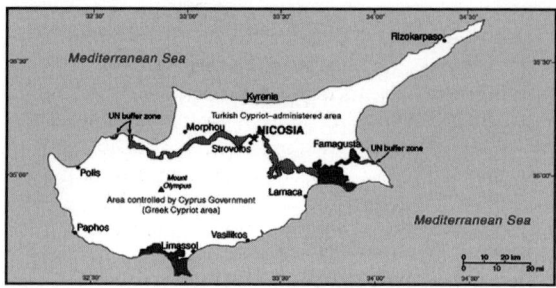

Cyprus, showing the division between government-controlled territory (5B) and the unofficial Turkish state of northern Cyprus.

(FRC). These activities include stations organised for international contests, HF or VHF / UHF DXpeditions, and special event stations organised by the FRC. Such contest activities, DXpeditions and special event stations are usually given callsigns with the special T4 prefix, rather than the more usual CO or CM.

The foreign amateurs must be on tourist packages which include hotels and all transportation to and from the working location. The FRC will apply for the temporary radio licences and any other authorisation necessary for the foreign amateurs to operate from Cuba and, in the case of contest stations and DXpeditions, for using their own callsigns/CO before and after the activity.

Visitors participating in radio activities organised by the FRC may import, free of duty and tax, all the equipment and antennas necessary for the activity. The FRC will obtain the necessary customs permission but the visiting amateurs must send a complete list of the equipment they will take with them at least 30 days before their arrival.

Any amateur wishing to operate in Cuba under these conditions should contact the FRC by e-mail to frcuba_en@enet.cu in order to make a proposal. The activity must be approved by the FRC's Board of Directors.

Licensing authority: Direccion de Frecuencias Radioelectricas, Ministerio de Informática y Comunicaciones, Avenida Independencia No 2 e/19 de Myo y Aranguren, Plaza de la Revolución, La Habana, CP 10600, Cuba. **Tel:** +53 7 885 4076 / 4080; **fax:** +53 7 881 2856.

CYPRUS (CEPT, HAREC)

If you are from a country with a CEPT licence you may operate for up to three months in Cyprus using the prefix 5B/ before your own callsign. Cyprus also accepts the HAREC, so you may apply for a full Cypriot licence if you plan to stay for longer than three months.

If you do not have a CEPT licence or HAREC, you can apply to the Department of Electronic Communications (DEC), a department of the Ministry of Communications and Works, either for a

short-term visitor's licence, or a permanent 5B callsign. An application form for an amateur radio licence, in Greek and English, can be found on the DEC website by clicking on "English", then "Authorisations" and "Application forms".

Licensing authority: Department of Electronic Communications (DEC), Ministry of Communications and Works, PO Box 24647, 1302 Nicosia, Cyprus; **website:** www.mcw.gov.cy/dec **Tel:** +357 22 814 840; **fax:** +357 22 321 925. **E-mail:** info.dec@mcw.gov.

CYPRUS – BRITISH SOVEREIGN BASE AREAS

Note that CEPT, HAREC and Cypriot-issued (5B) licences are not valid in the British Sovereign Base Areas of Cyprus. At present, only British citizens *resident* in Cyprus may obtain a ZC4 callsign (short-term visitors are not issued with a licence). However, in addition to the callsign it is also necessary to have permission to establish a transmitting station within the Sovereign Base Areas. This permission is normally only granted to British Forces personnel and staff working and or living within the British Sovereign Base Areas.

Licensing authority: HQ SBAA, Episkopi 3370, Cyprus; **website:** www.sba.mod.uk **Fax:** +357 25 963 521. **E-mail:** hqsbaa@cytanet.com.cy

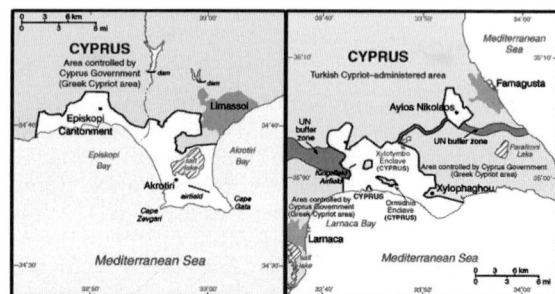

British Sovereign Base Areas on Cyprus, ZC4: West (left) and East (right).

CZECH REPUBLIC (CEPT, HAREC)

The Czech Republic is another country that has adopted both CEPT Recommendation T/R 61-01 and T/R 61-02, so licensing is not a problem for most amateurs. If you do not hold a CEPT licence or a HAREC, apply to the Amateur Radiocommunication Service of the Czech Telecommunications Office for a visitor's licence. The prefix for visitor's licences is OK8 and the licence fee is CZK 500 crowns (approx GBP £14.50).

Licensing authority: Amateur Radiocommunication Service, Department of Frequency Spectrum Management, Czech Telecommunications Office; **website:** www.ctu.cz

Postal address: Cesky telekomunikacni urad, PO Box 02, CZ-225 02 Praha 025, Czech Republic.
Street address: Cesky telekomunikacni urad, Sokolovska 219, Praha 9.
Tel: +420 224 004 725; **fax:** +420 224 004 823
E-mail: info@ctu.cz

DENMARK (CEPT, HAREC)

There is no problem to obtain an amateur radio licence in Denmark. If you do not hold a CEPT Licence or a HAREC, apply to NITA for a visitor's licence.

The same applies to licences for the Faroe Islands and Greenland.

Licensing authority: National IT and Telecom Agency (NITA), Holsteinsgade 63, DK-2100 Copenhagen, Denmark; **website:** www.itst.dk
Tel: +45 35 450 000; **fax:** +45 35 450 010.
E-mail: itst@itst.dk

DJIBOUTI

Licensing for visitors to Djibouti is carried out on a case-by-case basis. Visitors are recommended to contact the President of Association des Radioamateurs de Djibouti (ARAD), Mohamed Omar Moussa, J28AP, for assistance. The address is ARAD, PO Box 1076, Djibouti, tel: +253 35 2490; fax: +253 35 5757.

Licensing authority: Ministère de la Communication, de la Culture, des Postes et Télécommunications, BP 32, Djibouti, Djibouti.
Tel: +253 353 928; **fax:** +253 353 957.
E-mail: mccpt@intnet.dj

DOMINICA

A licence is easily obtained in the Commonwealth of Dominica and visitors are normally granted a licence with the prefix J79. An eight-page application form in pdf format is available on the website of the Dominica Amateur Radio Club Inc (DARCI) at www.j7hams.com/visitor.htm

Complete the application form and attach a copy of your current home licence and your passport. The application fee is USD $25.00 if you pay by cash. If you pay by cheque, there is a surcharge of $2.00. Cheques should be made out to the Ministry of Housing, Lands, Communications, Energy and Ports.

The Dominican licence can be picked up at any port of entry on the island upon your arrival if you inform the National Telecommunications Regulatory Commission of that port of entry and the date of your arrival. A copy of the licence will also be faxed to you upon request. In order to pick up the original licence at the port of entry, your payment should have been made prior to your arrival. If you do not intend to pick up your licence at the port of entry, it must be collected in person from the National Telecommunications Regulatory Commission office in Roseau.

The person in charge of licensing is Mr George A James.

Licensing authority: The Chairman, National Telecommunications Regulatory Commission, Ministry of Housing, Lands, Communications, Energy and Ports, Government Headquarters, 2nd Floor, 42-2 Kennedy Avenue, Roseau, Commonwealth of Dominica.
Tel: +1 767 440 0627 / +1 767 500 3333 / +1 767 448 2401;
fax: +1 767 440 0835 / +1 767 448 4807.
E-mail: gjames@ectel.int, secretariat@ntrcdom.org *or* codtelecom@cwdom.dm

DOMINICAN REPUBLIC

In order to operate amateur radio in the Dominican Republic, a visitor's licence is required. The guest licence is issued on the basis of reciprocal agreements by the Instituto Dominicano de las Telecomunicaciones, INDOTEL. If you will be in Santo Domingo you may apply for the licence in person, but if applying by post you should send the paperwork at least three to four weeks before your planned arrival date in the Dominican Republic.

The callsign issued is your home call *followed by* /HI followed by a call area digit, e.g. 3 for Santiago, 8 for Santo Domingo, 9 for Samaná (the Dominican Republic country identification should *not* be placed as a prefix before your own callsign.)

The INDOTEL website has information (in Spanish) about amateur radio, including a licence application form. The application form can be downloaded by clicking on "Medios Electrónicos", "Radioaficionados", "Solicitud y Renovación de Licencias para Radioaficionados", and "Formulario de Declaración".

Complete and sign the application form, write a separate letter of application in Spanish and enclose two 2 x 2in photographs and copies of your home licence and passport. The licence fee is DOP 100 pesos plus 25 pesos (total approx GBP

£2) if you wish to have the licence sent by post or fax to where you are staying. The application can be sent to INDOTEL by air mail, fax or e-mail.

Licensing authority: Instituto Dominicano de Telecomunicaciones (INDOTEL), Avenida Abraham Lincoln 962, Edificio Osiris, 2a Planta, Santo Domingo, Dominican Republic;
website: www.indotel.org.do
Tel: +1 809 473 8526; **fax:** +1 809 732 3904.
E-mail: info@indotel.org.do

EAST TIMOR – SEE TIMOR-LESTE

ECUADOR

If you wish to operate in Ecuador (including the Galapagos Islands), you should apply for a visitor's licence through one of the regional radio clubs. The application is forwarded by the radio club to the licensing authority. Write a letter in Spanish requesting a temporary visitor's licence and attach copies of your home licence and passport as well as two passport-type photographs. The licence fee is USD $30.

Two radio clubs that are known to have assisted overseas visitors to obtain Ecuadorian visitor's licences are the Quito Radio Club and the Guayaquil Radio Club. You are advised to use a courier service such as FedEx, DHL or UPS to send the documentation. Contact details are: Quito Radio Club, Calle Cochapata 224 y Jose Maria Abascal, Quito, Ecuador; tel: +593 2 225 5427, *or:* Guayaquil Radio Club, PO Box 09-01-5757, Guayaquil, Ecuador; tel: +593 4 294671 / +593 4 294682.

Juan Jose Forestieri, HC2FN, former president of the Guayaquil Radio Club, has also helped visitors to obtain a licence. Contact him by e-mail: jjfores@hotmail.com

Licensing authority: Consejo Nacional de Telecomunicaciones (CONATEL), Edificio Zeus, Av Diego de Almagro N 31-95 y Alpallana, Casilla 17-07-9777, Quito, Ecuador;
website: www.conatel.gov.ec
Tel: +593 2 222 5614;
fax: +593 2 250 5119 / +593 2 223 1591.
E-mail: presidencia@conatel.gov.ec

EGYPT

Generally, it is not possible to obtain a short-term visitor's licence in Egypt. For limited duration operations, it is possible for visitors to operate as a guest from existing licensed Egyptians' stations using their callsign.

However, if you are resident in Egypt, for example with a work permit, you should be able to obtain an SU9 licence. It is recommended that you apply through the Egypt Amateur Radio Assembly (EARA); website: www.qsl.net/egyptham

The Secretary of EARA is Mohammed Al-Kafrawi, SU1KM, PO Box 70, Magless, El Shabb, Cairo, 11516 Egypt; mob: +20 12 210 6000; fax: +20 2 3836 9850; e-mail: su1km@egyptham.net or su1km@link.net

Licensing authority: National Telecommunication Regulatory Authority (NTRA), Building B4, Smart Village, Km 28, Cairo-Alexandria Desert Road, Cairo, Egypt; **website:** www.ntra.gov.eg
Tel: +20 2 534 4124; **fax:** +20 2 534 4155 / 6.
E-mail: info@ntra.gov.eg

EL SALVADOR (IARP)

The easiest way to operate in El Salvador is if you can obtain an IARP. If you do not hold one, it is recommended that you apply for an El Salvador visitor's licence through the country's national society, Club de Radio Aficionados de El Salvador (CRAS), PO Box 517, San Salvador, El Salvador; tel: +503 248 3905; fax: +503 248 3906; e-mail: cras@intercom.sv *or* cras_elsalvador@yahoo.com The HQ is located at Calle 1, No 5, Urbanizacion Lomas de San Francisco, San Salvador.

Licensing authority: Superintendencia General de Electricidad y Telecomunicaciones (SIGET), Sexta Décima Calle Poniente y 37 Av Sur, No 2001, Col Flor Blanca, San Salvador, El Salvador;
website: www.siget.gob.sv
Tel: +503 2257 4401 / 4484;
fax: +503 2257 4493 / 4498.
E-mail: siget@siget.gob.sv

EQUATORIAL GUINEA

No up to date information available. Visitors' licences are issued from time to time but it is almost certainly easier to obtain one in person, rather than attempting to apply by post or fax.

Licensing authority: Ministerio de Transportes, Tecnología, Correos y Telecomunicaciones, Dirección General de Correos y Telecomunicaciones, Malabo, Equatorial Guinea.
Tel: +240 093 029 / +240 098 756;
fax: +240 092 618 / +240 093 202.

ERITREA

It is sometimes possible, sometimes impossible, to obtain an amateur radio licence in Eritrea (presumably the availability, or otherwise, of licences is dependent upon the prevailing political and military situation in the country at the time).

If and when licences are being issued, the licence fee is USD $500 *per person* for a one-year licence, and each person who wishes to operate *must* have their own individual licence (so, for example, the licence fee for a DXpedition team of six operators would be $3000).

One person who has issued amateur radio licences in the past, Mr Zerai Teklehaimanot, is now (2008) the Director of the Communications Department at the Ministry of Transport and Communications and it is suggested that applications for an amateur licence would best be directed personally to him. An application form for a "Radio Transmitter Permit" is available on the Internet at www.qsl.net/oh2mcn/e3a.htm In addition to the Radio Transmitter Permit application form you should submit a letter requesting permission to operate, and provide full technical specifications (in English) of the equipment and antennas to be used. The licence and callsign are only issued after your arrival in the country and the licence document must be stamped before any operation takes place.

Note that the Ministry of Transport and Communications has several PO Box numbers in Asmara. Both PO Box 4918 and PO Box 5417 may be used for the Communications Department.

Licensing authority: Ministry of Transport and Communications, Communications Department, Harnet Street, PO Box 5417, Asmara, Eritrea.
Tel: +291 1 126 965 / +291 1 115 847;
fax: +291 1 126 966.
E-mail: zerait@eol.com.er

ESTONIA (CEPT, HAREC)

Both the CEPT licence and the HAREC are accepted in Estonia. Note, however, that confirmation of Morse code ability at a minimum of 5 words per minute is required for HF bands operation.

If you do not have a CEPT licence or a HAREC you will need to apply for a visitor's licence in order to operate in Estonia. The licensing authority is the TJA (Tehnilise Järelevalve Amet in Estonian or Estonian Technical Surveillance Authority, ETSA, in English). TJA was created in 2008 from the former Estonian National Communications Board as well as the Estonian Railway Inspectorate and Estonian Technical Inspectorate.

TJA has a website in Estonian, English and Russian. Click on "ENG" at the top right of the screen, then "Division of Electronic Communications" then "Radio communications" for several useful pages about amateur radio licensing in Estonia.

Application for a visitor's licence should be made to Ms Kristiina Kenk at the Frequency Management Department of TJA at the address below. The licence fee is EEK 36.00 krooni (Estonian crowns, approx GBP £1.75) for a period of up to three years.

Licensing authority: Frequency Management Department, Tehnilise Järelevalve Amet (TJA), Sõle 23 A2, 10614 Tallinn, Estonia;
website: www.tja.ee
Tel: +372 667 2000 / 2130; **fax:** +372 667 2001.
E-mail: kristiina.kenk@tja.ee or info@tja.ee

ETHIOPIA

It is quite easy to get an amateur radio licence in Ethiopia if you have a residence permit, but more difficult for short-term visitors to the country. Amateur radio licensing is carried out by Ethiopian Telecommunication Agency (ETA) and Annex III on their website gives details of how to apply for a licence - click on "Licenses", "Requirements (to acquire ETA's service)", then "Radio Communication" to download the pdf. The requirements are: 1) A passport with a copy of the first four pages of the passport; 2) A residential permit; 3) Amateur radio communication station licence from other country, if any; 4) One portrait photo of the applicant taken within the last six months; and 5) Written application letter for amateur licence to the Agency. The "appropriate radio frequency spectrum fee" must also be paid.

An official application form, also in pdf format, is available by clicking on "Licenses", "License Application Forms" then "Radio Communication License Application Form".

Licensing authority: Licensing and Inspection Directorate, Ethiopian Telecommunication Agency (ETA); **website:** www.eta.gov.et
Postal address: PO Box 9991, Addis Ababa, Ethiopia.
Street address: 2nd Floor, Bekelobet, Tegene Building, Kirkos District, K02/03, House No 542, Addis Ababa.
Tel: +251 11 466 8203 / 8282;
fax: +251 11 465 5763.
E-mail: frequency@eta.gov.et *or* tele.agency@ethionet.et

FALKLAND ISLANDS

VP8 amateur radio licences for the Falkland Islands, as well as South Georgia, the South Sandwich Islands, and British bases on Antarctica, the South Shetland Islands and the South Orkney Islands, can be obtained from the Post Office in Stanley, Falkland Islands.

Either apply in person or you can phone or e-mail for an application form which will be faxed to you. When the form is completed it can be sent by air mail post to the Falklands. The licence fee is GBP £20 for a one-year licence.

Note that the VP8 licence is normally valid for use in the Falkland Islands only. If you require a licence for the other VP8 territories, this must be specifically requested. In addition to the licence, separate permission is required to visit South Georgia and the South Sandwich Islands.

Licensing authority: Superintendent of Posts and Telecomms, c/o Post Office, Stanley, Falkland Islands, South Atlantic;
Tel: +500 27180.
E-mail: adminpost@townhall.gov.fk

FAROE ISLANDS – SEE DENMARK

FEDERATED STATES OF MICRONESIA – SEE MICRONESIA

FIJI

If you intend taking any transmitting equipment into Fiji (including VHF or UHF handhelds) it is necessary to have an import permit *as well as* a licence. If you are bringing a transmitter into Fiji you must go through the Red Channel at customs and declare the equipment. Even if you have a licence, the equipment will be impounded unless you also have the import permit. Fortunately, it is possible to apply for the licence and import permit at the same time. The fee is USD $15 (less if an import permit is not required).

Apply by letter listing your home address, telephone number, e-mail address and fax number (if available) and state the location of operation and the period of operation in Fiji. Enclose a copy of your home licence, one passport-type photograph and the licence fee. If an import permit is required this should be specifically requested and the make, model and serial number of each piece of equipment must be listed. You may only import the equipment listed on the import permit so you should ensure that the serial numbers are listed correctly.

The licensing officers are Jone Buliruarua and Jale Curuki. You may request a specific 3D2 two-letter callsign, but most have already been used and many calls are re-issued, so give a choice of three or four possible callsigns. The licence and import permit will be posted to your home address if the application was received sufficiently early; it is recommended that you allow at least three months for the applications to be processed before your arrival in Fiji.

Note about operating from Rotuma Island: Licensing for Rotuma is handled by the Department of Communications office in Suva, but you need permission from the island's chiefs to stay on the island. There are no tourist facilities on Rotuma so it is necessary to stay with a family. Facilities are very basic. Aisea Aisake, 3D2AA, is a Rotuman resident in Lautoka on the main Fijian island of Viti Levu and he may be able to arrange permission and accommodation for adventurous amateurs wishing to operate from Rotuma. Contact him by e-mail: aisea@fsc.com.fj

Licensing authority: Radio Regulatory Unit, Department of Communications, Ministry of Industry, Tourism, Trade and Communications; **website:**
www.fiji.gov.fj/publish/page_3514.shtml
Postal address: Government Buildings, PO Box 2118, Suva, Fiji.

Street address: Department of Communications, Credit Corporations Building, 91 Gordon Street, Suva.
Tel: +679 331 8507 / +679 338 4766;
fax: +679 338 6310 / +679 331 5167.
E-mail: jcuruki@connect.com.fj *or* jbuliruarua@connect.com.fj *or* jturaganivalu@connect.com.fj

FINLAND (CEPT, HAREC)

If you have a CEPT licence, you may operate in Finland without notifying the Finnish authorities in advance. If you plan to stay in Finland for longer than three months, or make repeated visits, you may also apply for a permanent Finnish amateur licence on the basis of a HAREC.

The website of the Finnish Communications Regulatory Authority (FICORA), in Finnish, Swedish and English, has a useful section about amateur radio regulations and licensing in Finland.

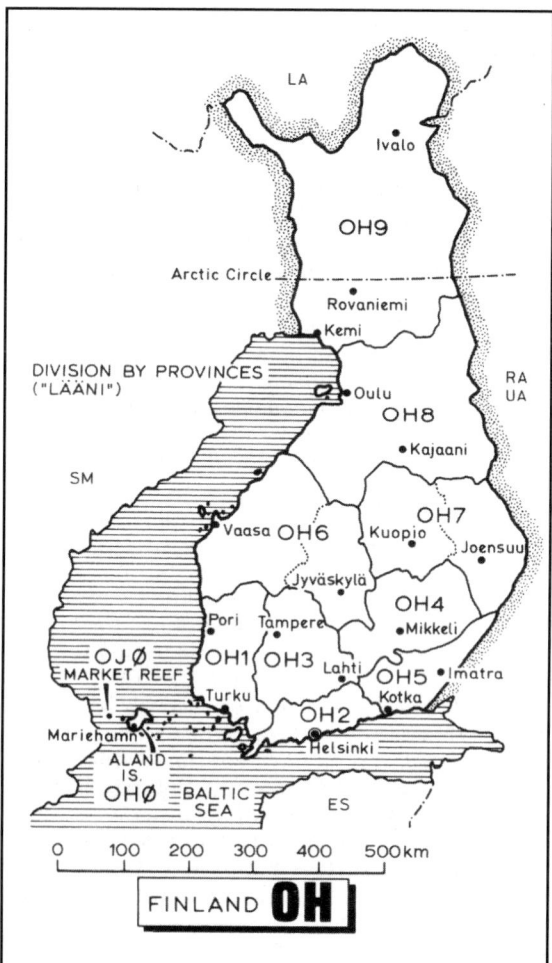

Finland, OH, call districts (see text).

You may download an application form for a permanent licence from the website.

If you have neither a CEPT licence nor a HAREC, you should apply for a short-term visitor's licence. There is an application form at www.ficora.fi/englanti/lomake/ATUe.pdf

Finally, it is possible to obtain a permanent Finnish licence by taking the Finnish amateur radio examinations, which are available in English as well as Finnish and Swedish. (An English-language exam must be requested two months before the date of the exam. Contact the Finnish national amateur radio society, SRAL, if you wish to take this option.)

Although Finland still has a call district system, 'vanity' callsigns may use any digit. The exception is OH0, which is reserved for operation from the Åland Islands only.

Licensing authority: Finnish Communications Regulatory Authority (FICORA), Itämerenkatu 3A, PO Box 313, FI-00181 Helsinki, Finland;
website: www.ficora.fi
Tel: +358 9 6966 436 / 479 / 482 / 883;
fax: +358 9 6966 410.
E-mail: radiocom@ficora.fi

FRANCE (CEPT)
Radio licensing in France is divided between ARCEP (Autorité de Régulation des Communications Électroniques et des Postes, www.arcep.fr/index.php?id=8138), which provides the regulatory framework, and ANFR (Agence Nationale des Frequences), which coordinates frequency usage and issues licences.

An application form ("Demande d'indicatif pour radioamateur etranger etabli en France") intended for foreign amateurs staying in France for longer than three months (who are issued with F5Vxx callsigns) is available on the ANFR website at www.anfr.fr/index.php?cat=radioamateur

According to Appendix II of CEPT Recommendation T/R 61-01, the CEPT Licence is also valid in the following French territories: Corsica, Guadeloupe, Guyana, Martinique, St-Bartholomew, St-Pierre / Miquelon, St-Martin, Réunion (Glorieuse, Jean de Nova [sic], Tromelin), Mayotte, French Antarctica (Crozet, Kerguelen, St Paul & Amsterdam, Terre Adelie), French Polynesia & Clipperton [but see below], New Caledonia, Wallis & Futuna. Although the CEPT Licence provides authority to operate amateur radio, it is still necessary to obtain permission to visit some of them (e.g. Glorieuses, Juan de Nova and Tromelin).

Licensing authority (head office): Agence Nationale des Fréquences (ANFR), 78 Avenue du Général de Gaulle, 94704 Maisons-Alfort, France;
website: www.anfr.fr
Tel: +33 1 4518 7272.

FRENCH POLYNESIA (CEPT)
The latest version of Appendix II of CEPT Recommendation T/R 61-01 lists French Polynesia & Clipperton as one of the French territories in which the CEPT Licence is valid. Morse proficiency is required for the use of HF. However, until recently it was necessary to apply for a licence locally in order to operate amateur radio from French Polynesia and it is believed that the PTT in Papeete, Tahiti, does still issue licences locally. (This could therefore be a case of the authorities in mainland France not being aware of the requirements of the authorities in Tahiti, or the Tahitian authorities not being willing to accept directions from Paris. It is recommended that local advice be taken in Tahiti and, if necessary, to apply for a licence there.)

An application form in French and English is available on the Internet at www.qsl.net/oh2mcn/fo8a.jpg Documents required are a *certified* copy of your amateur radio licence and your passport, as well as a list of all equipment to be used in French Polynesia, including serial numbers. Note that only licences equivalent to CEPT Class 1 (i.e. with Morse code qualification) are accepted. The FO0 visitor's licence is issued free of charge for a period of up to 90 days.

If you apply in person you should go to the Cellule des Postes et Télécommunications ('Celle PTT') office in Papeete. The correct Celle PTT office is close to the large Gendarmarie (police) compound on Avenue Bruat.

Licensing authority: Cellule des Postes et Télécommunications du Haut Commissariat de la Republique (HCR-CPT), BP 115, 98713 Papeete, Tahiti, French Polynesia.
Tel: +689 468 630 / 632; **fax:** +689 468 633.
E-mail: her.ept.jde@mail.pf

GABON
There is an application form in French for an amateur radio licence on the website of the Gabonese licensing authority, ARTEL, at www.artel.ga/formulaire.htm Four passport-type photographs are required with the application.

The IARU society, Association Gabonaise des Radio-Amateurs [AGRA], may be able to help with the application if you run into any difficulties. Their address is BP 1826, Libreville, Gabon, or contact the President of AGRA, Alain Combelles, TR8CA, BP 1293, Libreville, Gabon; tel: +241 730 154; fax: +241 702 425; e-mail: tr8ca@hotmail.com
Licensing authority: Agence de Régulation des Télécommunications (ARTEL);
website: www.artel.ga
Postal address: BP 50 000, Libreville, Gabon.
Street address: 413 Boulevard du Bord de Mer, Libreville.

Tel: +241 768 215;
fax: +241 765 746 / +241 721 725.
E-mail: artel@inet.ga

GAMBIA

Licensing in Gambia is relatively straightforward providing you go to the licensing office in person. Almost any vacant callsign is possible, with the prefix being either C5 or C5 followed by any other digit, so go with a few ideas!

The licence fee is GMD 1000 Gambian dalasi (approx GBP £24) for a one-year licence, but the biggest difficulty is in actually paying for the licence. A recent change in regulations means that only registered Gambian income tax payers can pay the fee. This means that foreign visitors must make arrangements with a resident to pay the licence fee on their behalf.

The contact person in 2007 was Mr Yankouba Toure but the 'man in charge' seems to change frequently in Gambia.
Licensing authority: Department of State for Communications, Information and Technology (DOSCIT), GTRS Building, MDI Road, Kanfing, The Gambia; **website:** www.doscit.gm
Tel: +220 437 8031; **fax:** +220 437 8029.
E-mail: doscit@gamtel.gm

GEORGIA

The Georgian National Communications Commission (GNCC) was established in 2000 and in 2007 Georgia became a full member of CEPT (the European Conference of Postal and Telecommunications Administrations). The GNCC collaborates closely with CEPT as the regulatory authority for telecommunications in Georgia. Unfortunately, Georgia has not yet adopted CEPT Recommendation T/R 61-01, so for the time being at least it is still necessary to apply to GNCC for a visitor's licence.
Licensing authority: Radio Frequencies Management, Monitoring and Coordination Department, Georgian National Communications Commission (GNCC), 42 A Kazbegi Avenue, Tbilisi 0177, Georgia; **website:** www.gncc.ge
Tel: +995 32 399 561 or +995 32 921 667;
fax: +995 32 921 625
E-mail: sshavgulidze@gncc.ge

GERMANY (CEPT, HAREC)

If you have a CEPT Licence you can operate in Germany without any further formalities. Likewise, you can apply for a full German licence if you have a HAREC and you plan to stay for more than three months.

Foreign radio amateurs without either of these should apply to the Mülheim regional office of the Bundesnetzagentur (BnetzA, or Federal Network Agency), which can issue licences with a validity period of up to three months. This regional office will send you a special information sheet and an application form. The licence fee is EUR 55.00 euros (approx GBP £45).

The BnetzA website has information in English about amateur radio licensing for visitors to Germany: go to www.bundesnetzagentur.de, click on "English" at the top right of the page, then "Telecommunications", and then finally "Amateur radio".

If you have residency in Germany, you can also apply for a full German licence (period greater than three months) whether or not you are from a HAREC-issuing country. You should submit a copy of your work permit or residence visa with the other documents and send the application to the regulatory authority's regional office responsible for your place of residence. You may request a specific callsign which will be issued if it is available: list a few possible calls.
Licensing authority: Bundesnetzagentur, ASt Mülheim, Postfach 10 03 51, 45403 Mülheim, Germany; **website:** www.bundesnetzagentur.de
Tel: +49 251 6081 280 / 281;
fax: +49 251 6081 180 (BNetzA HQ in Münster).
E-mail: MLHM01.Postfach@bnetza.de

GHANA

It is possible for foreign visitors to receive an amateur radio licence in Ghana. The 9G5 prefix is issued to non-residents. The licence fee is approximately USD $35 (less for renewals) and should be paid by banker's draft.

The Senior Manager, Regulations and Licensing at the NCA was Mrs Golda Adjei in early 2008.
Licensing authority: National Communications Authority (NCA), 1st Rangoon Close, PO Box CT 1568 Cantonments, Accra, Ghana.
Tel: +233 21 776 621; **fax:** +233 21 763 449.
E-mail: nca@ghana.com or golda.adjei@nca.org.gh

GIBRALTAR

Amateur radio licensing in Gibraltar is much more difficult than it used to be. Since June 2006 the Gibraltar Regulatory Authority (GRA) has suspended the issue of reciprocal licences for HF operation and visitors are now allowed the use of 6m and 2m only and with a maximum *ERP* of 100 watts (the use of linear amplifiers is expressly forbidden). This course of action was reportedly due to instances of visitors causing serious interference with essential services.

Gibraltar is not currently part of the CEPT scheme and foreign licences are not valid in the territory. The police and customs take a serious view of unlicensed operators and under Gibraltar law wireless equipment may be confiscated even if it is not actually in use.

If you wish to apply for a VHF licence, an information sheet and application form can be found on the GRA website under "Communications" and "Licences". There is no charge.

Licensing authority: Gibraltar Regulatory Authority (GRA), Suite 811, Europort, Gibraltar;
website: www.gra.gi
Tel: +350 200 74636; **fax:** +350 200 72166
E-mail: communications@gra.gi

GREECE (CEPT, HAREC)

Greece recognises both the CEPT Licence and the HAREC, but requires confirmation of Morse code ability at a minimum speed of 5 words per minute for HF bands operation.

Those from non-CEPT countries, or anyone staying for longer than three months in Greece, should apply to the Ministry of Transport and Communications for a 'special Greek radio amateur licence' with an SV0 callsign. In either case, you should apply in English (or Greek) with the following information:
• Application with personal ID details including national callsign, home address and address while in Greece;
• A photocopy of national radio amateur operating licence, indicating operating privileges, i.e. frequencies, power limit etc;
• A statement containing the permanent address in Greece;
• The accompanying fee (check with the ministry first).

The licensing officer is Mrs Christina Papathanasiou.

Note about operating from Mt Athos (SV/A or SY): Operation within the district of Mt Athos (also known as Agion Oros or the Holy Mountain) is subject to receiving official *written* permission of the local administration of the holy community. If you receive such permission you must then apply to the Ministry of Telecommunications for a licence. Both permissions are required for a valid operation.

Licensing authority: Ministry of Transport and Communications, Department of Communication Control, 2 Anastaseos Street and Tsigante, 101 91 Holargos, Athens, Greece;
website: www.yme.gov.gr
Tel: +30 210 650 8000 / 8555;
fax: +30 210 650 8460.
E-mail:
c.papathanasioy@yme.gov.gr

GREENLAND – SEE DENMARK

GRENADA

An application form for an amateur radio licence can be downloaded from Grenada's NTRC website.

Note that *two* copies of the completed application form should be submitted to the NTRC, accompanied by identification (e.g. passport), a passport-size photo, the applicable fees, technical specification of each piece of equipment, and copy of your home licence.

The licence fee is XCD $50 Eastern Caribbean dollars (approx GBP £9.50) plus an initial application fee of EC$25 (approx GBP £4.75).

Licensing authority: National Telecommunications Regulatory Commission (NTRC)
Postal address: PO Box 854, St George's, Grenada;
Street address: Suite 8, Grand Anse Shopping Centre, Grand Anse, St George's;

Greece, SV, call districts.

website: www.ectel.int/grd
Tel: +1 473 435 6872; **fax:** +1 473 435 2132.
E-mail: gntrc@spiceisle.com

GUATEMALA

It should not prove to be too onerous to obtain a permit to operate amateur radio in Guatemala. Specifically, Guatemala has a reciprocal licensing agreement with USA. There is no fee.

The website of SIT, the licensing authority, includes a list of amateur radio frequencies, but no information about licensing in the country.
Licensing authority: Superintendencia de Telecomunicaciones de Guatemala (SIT), Edificio Murano Center, Nivel 16, 14 Calle 3-51, Zona 10, Guatemala City 01010, Guatemala;
website: www.sit.gob.gt
Tel: +502 2366 5890; **fax:** +502 2366 5880.

GUINEA

It is possible to obtain an amateur radio licence in Guinea by applying in person to the Ministry of Communications. Callsigns normally issued to visitors are in the series 3XY followed by a single digit and one or two letters (the ITU-allocated prefix is 3X, the 'Y' indicates a station in the Amateur Service, and the following digit and letter are issued in numerical and alphabetical order).

The National Director of Posts and Telecommunications is Mr Mamadou Dioulдé Sow.
Licensing authority: Ministère de la Communication et des Nouvelles Technologies de l'Information, BP 3000, Conakry, Republic of Guinea.
Tel: +224 30 454 586 / +224 30 435 500;
fax: +224 30 451 896.
E-mail: msow@sotelgui.net.gn

GUINEA-BISSAU

It is possible to obtain an amateur radio licence in Guinea-Bissau. Apply in person to the ICGB.
Licensing authority: Instituto das Comunicações da Guiné-Bissau (ICGB), R Vitorino Costa, CP 1372, Bissau, Republic of Guinea-Bissau;
website: www.icgb.org
Tel: +245 20 4873 /4874; **fax:** +245 20 4876.
E-mail: icgb@mail.bissau.net

GUYANA

Guyana only has reciprocal licensing agreements with two countries - the USA and Trinidad and Tobago - and a visitor's licence will only be issued to foreigners holding licences issued in either of those countries. If you do not have either of these the only other way to get a licence in Guyana is to sit and pass the local examinations. Exams are set at three different levels corresponding with the three classes of licence: Technician,

General and Extra.

The licensing authority in Guyana is the National Frequency Management Unit (NFMU). To apply for a licence, first download and print off an application form which can be found on the Internet by following the links from the Guyana page from OH2MCN's site at www.qsl.net/oh2mcn/license.htm (the application form is not available on the NFMU's own website). The original hard copy of the application form must be submitted to the NFMU office (not e-mailed) along with a processing fee of GYD $500 Guyana dollars (approx GBP £1.25).

It is necessary to receive permission from the NFMU before importing transmitting or receiving equipment into Guyana. You need to fill out an 'Import License' application form and submit it to the NFMU for a "no objection to importation" stamp. The 'Import License' application form can be obtained from the Ministry of Tourism, Industry & Commerce or at local stationery stores in Guyana. The stamped form should then be submitted to the Ministry of Tourism, Industry & Commerce for processing.
Licensing authority: Mr Valmikki Singh, Acting Chief Executive Officer, National Frequency Management Unit (NFMU); **website:** www.nfmu.gov.gy
Postal address: NFMU, PO Box 12174, Georgetown, Guyana.
Street address: NFMU, 68 Hadfield Street - Lodge, D'Urban Park, Georgetown.
Tel: +592 226 2233 / 3976 / +592 225 3104;
fax: +592 226 7661
E-mail: nfmu@sdnp.org.gy

HAITI

A visitor's licence of up to three months duration is available in Haiti. Apply in person with the usual documentation plus two passport photographs. You will be required to fill in some application forms so you are advised to take a translator if you do not read or write French.
Licensing authority: Conseil National des Télécommunications (CONATEL), 16 Cité de l'Exposition, BP 2002, Port-au-Prince, Haiti;
website: www.conatel.gouv.ht
Tel: +509 2222 0300/ 2221 8305 / 2223 0720;
fax: +509 2223 0579 / +509 2223 9229.
E-mail: info@conatel.gouv.ht

HONDURAS

It will take about three months to obtain a Honduran licence with HR or HQ callsign. However, a short-term visitor's permit with your home call/HR can be organised through the IARU member society, the Radio Club de Honduras, in about two weeks.

The following information is required: your

full name, address in home country, telephone and fax number, e-mail address, the address where you will be staying in Honduras and/or contact name (if not staying at an hotel), phone and fax numbers at the place you are staying, e-mail address where you are staying, your arrival and departure dates in the country, and the equipment, bands and operating modes you intend to use.

Attach copies of your home licence, copy of passport personal details page(s) and Honduran visa page (if applicable), and send the whole application to CONATEL by registered mail, with a copy of everything to the Radio Club de Honduras, PO Box 273, San Pedro Sula, Cortés, Honduras (tel: +504 556 6173, +504 3392 6687 or +504 9818 8184; e-mail: hr2rch@yahoo.com; or contact Pedro, HR2PAC, e-mail: hr2pac@yahoo.com or Arturo, HR2AAB, e-mail: arturoaleman@amnethn.com)

If you are applying for a full HR or HQ callsign you may request a special callsign if you wish, e.g. for a contest, DXpedition or special event station. Do not send any money at the time of your application.

Licensing authority: Comisión Nacional de Telecomunicaciones (CONATEL), Edificio CONATEL

Postal address: PO Box 15012, Comayagüela, Honduras.

Street address: 6th Avenue Southwest, Colonia Modelo, Comayagüela.

website: www.conatel.hn
Tel: +504 234 8600; **fax:** +504 234 8611.
E-mail: secretaria@conatel.hn *or* conatel@conatel.hn

HONG KONG (HAREC)

It is possible to get a visitor's licence in Hong Kong without too much difficulty. Hong Kong accepts the City & Guilds of London Radio Amateur Examination Certificate, the HAREC and licences issued by other administrations if they are recognised by the Office of the Telecommunications Authority as equivalent to the Hong Kong licence.

The applicant should submit photocopies of their passport, amateur radio certificate (or licence) and documentary evidence issued by recognised organisations showing a pass in a Morse code test at either 12WPM or 5WPM for an HF licence. Fill in application form 'OFTA A201', which is available on OFTA's website and send the application to OFTA. Applications may also be made on-line via the website under "OFTA Services / Amateur Radio Services / How to Apply".

The licence is valid for one year and costs HKD $150 Hong Kong dollars (approx GBP £9.80).

Licensing authority: The Licensing Unit, Enforcement Section, Office of the Telecommunications Authority (OFTA), 29/F Wu Chung House, 213 Queen's Road East, Wan Chai, Hong Kong, SAR China; **website:** www.ofta.gov.hk
Tel: +852 2961 6752; **fax:** +852 2838 5112.
E-mail: ama_enq@ofta.gov.hk

HUNGARY (CEPT)

Licensing in Hungary is carried out by the Hungarian National Communications Authority (Nemzeti Hírközlési Hatóság, NHH, in Hungarian).

If you hold a CEPT licence you may operate in Hungary without further ado. If not, apply to the NHH for the three-month visitor's licence. Morse code is no longer a requirement for access to the HF bands, but the CW mode can only be used if the licensee has passed a Morse proficiency exam. The licence fee is HUF 2200 Hungarian forints (approx GBP £6.70), which is payable in tax stamps ("illetekbelyeg") available from post offices in Hungary.

Licensing authority: Hungarian National Communications Authority (NHH), Ostrom u 23-25, PO Box 75, 1525 Budapest, Hungary; **website:** www.nhh.hu
Tel: +36 1 457 7100 / 7488; **fax:** +36 1 356 5520.

Hungary, HA / HG, call districts (prefixes and letter call series).

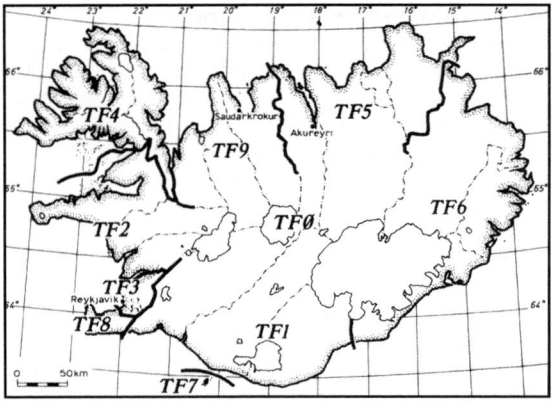

Iceland, TF, call districts.

ICELAND (CEPT, HAREC)

Iceland recognises both the CEPT and HAREC licences but if you do not have either you can still apply for a visitor's licence. An "Application form for a radioamateur license" is available on the PTA website: click on the British flag at the top right of the home page for the English-language website, then on "Applications and forms". You are advised that you do need to receive a statement or comment from your national IARU society as incomplete applications are rejected.

Licensing authority: Post and Telecom Administration, Suðurlandsbraut 4, 108 Reykjavik, Iceland; **website:** www.pta.is
Tel: +354 510 1500; **fax:** +354 510 1509.
E-mail: pta@pta.is

INDIA

It is possible to get a visitor's licence in India, but it takes a good deal of time and dedication to obtain one independently. An easier option is to obtain a licence for one of the occasional 'Hamfest' celebrations organised by India's National Institute of Amateur Radio (NIAR), e.g. the January 2007 VU7RG operation from the Lakshadweep Islands and the NIAR Silver Jubilee celebrations in October / November 2008.

An "Application for the Grant of an Indian Amateur W/T Licence to Foreign National" is available on NIAR's website at www.niar.org/sj/index.htm No fewer than seven copies of the form should be completed. You will need to provide particulars of the transmitting and receiving equipment and antennas to be used, the frequency bands and types of modulation to be used, and two passport-type photographs signed and dated on the front.

If you need further assistance, NIAR may be contacted as follows: The Administrative Officer, National Institute of Amateur Radio, Raj Bhavan Road, Hyderabad 500 082, India; tel: +91 40 2331 0287; e-mail: info@niar.org *or* niarindia@hotmail.com

A 34-page pdf document called 'The Indian Wireless Telegraphs (Amateur Service) Rules, 1978' is available for downloading from the VU7RG DXpedition website at www.dx-pedition.de/lakshadweep2007/vu-license-details.pdf The document also includes various licence application forms for Indian nationals.

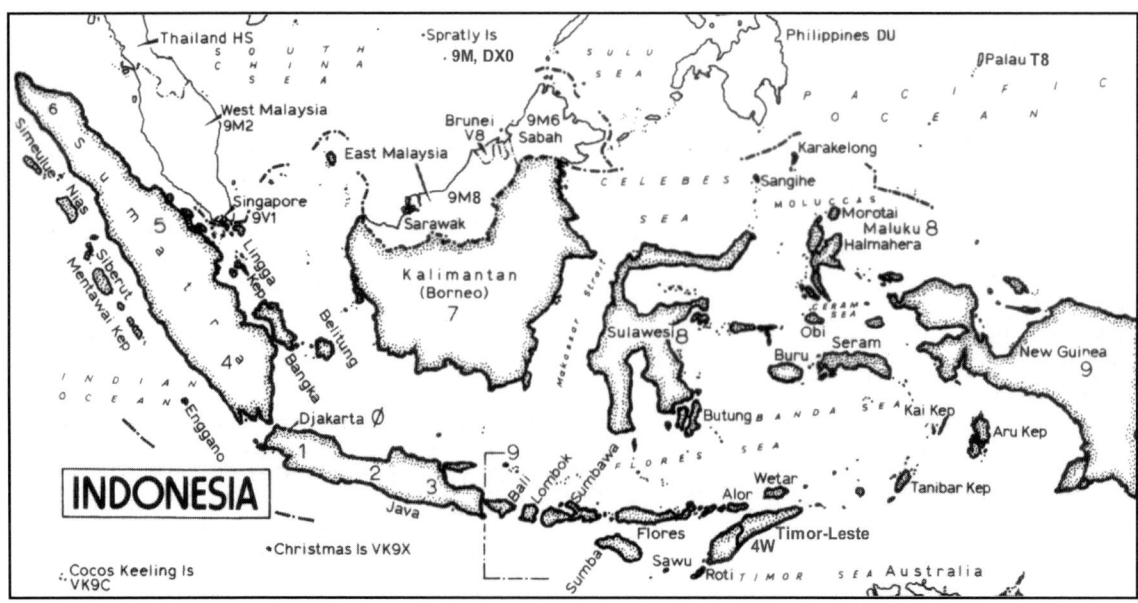

Indonesia, YB, call districts and surrounding DXCC entities.

Licensing authority: The Asst Wireless Advisor to Govt of India, Ministry of Communication & IT, Department of Telecommunications, WPC Wing, Amateur Section, 6th Floor, 521 Sanchar Bhavan, 20 Ashoka Road, New Delhi 110001, India.; **website:** www.wpc.dot.gov.in
Tel: +91 11 2303 6181; **fax:** +91 011 2371 6111.

INDONESIA

Foreigners need a permanent Indonesian address (not just a hotel address) before a licence will be granted. Applications should be made via the national society ORARI: you are advised to contact an English-speaking Indonesian amateur for help with this. Short-term visitors are granted a YBn/own call licence; if you have a one-year residence permit you can apply for a full YBnAxx licence.

There are several application forms to complete, and all are only in the Bahasa Indonesia language. The application is likely to take about three months, perhaps less in Jakarta. There are various fees to pay, and stamp duty on each application form. The total licence fee is approximately IDR 250,000 Indonesian rupiahs (approx GBP £14) for a one-year licence.
Licensing authority: Directorate General of Posts and Telecommunications (POSTEL), Ministry of Communication and Information Technology, Jalan Medan Merdeka Barat 17, 10110 Jakarta, Indonesia; **website:** www.postel.go.id
Tel: +62 21 345 6332 / +62 21 383 5888;
fax: +62 21 386 0746 / 0754 / 0781.

IRAN

At present it is virtually impossible for visitors to get an amateur radio licence in Iran. Foreigners temporarily resident in the country, for example diplomats, United Nations employees or those with work permits, who wish to apply for a licence are advised to contact the Chairman of the Amateur Radio Society of Iran, Mr Abdollah Sadjadian,

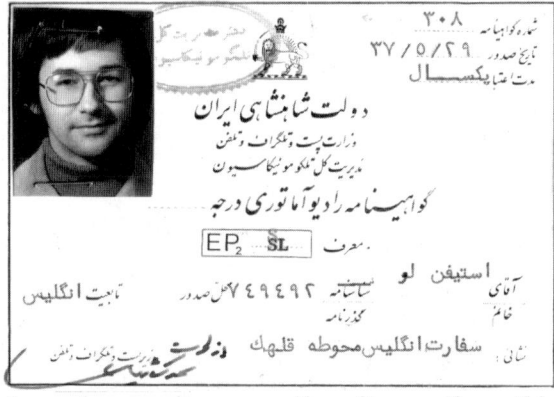

A much rarer item now than it was then, this is the author's EP2SL licence dating from 1978.

EP2FM, by e-mail at as@neda.net
Licensing authority: Communications Regulatory Authority (CRA), Ministry of Information and Communication Technology, PO Box 15875-4415, 15598 Tehran, Islamic Republic of Iran.
Tel: +98 21 8840 3612 / +98 21 811 2995;
fax: +98 21 8846 8999 / +98 21 860 0070.
E-mail: irnadm@cra.ir

IRAQ

Due to the security situation in Iraq, all amateur radio operation was prohibited during 2007, but on 20 November 2007 previously licensed amateurs were allowed to resume operations. New licences have since been issued.

All applications should go through the Iraqi Amateur Radio Society (IARS), PO Box 55072, 12001 Baghdad, Iraq; e-mail: yi1irq@yahoo.com The Society assesses the application and forwards it to the National Communications and Media Commission (CMC). A CMC application form is available on the IARS website at www.iraqi-ars.org In addition to the completed application form, you should provide a copy of your home licence and three passport-type photographs.
Licensing authority: Iraqi National Communications and Media Commission (CMC), Al-Masbah Mahala 929, Baghdad, Iraq.
Tel: +964 790 142 7265; **fax:** +964 1719 5839.
E-mail: amateurlicense@cmc.iq

IRELAND (CEPT, HAREC)

Radio amateurs are officially known as 'Radio Experimenters' in the Republic of Ireland, and the Irish Commission for Communications Regulation (CCR) has a good website with plenty of useful information for those wishing to operate from Ireland. Ireland accepts the CEPT licence for visits of up to three months and the HAREC for longer term licences. If you have neither, you may apply for a visitor's licence. The CCR has an application form in pdf format by going to www.comreg.ie then clicking on "Radio Spectrum", "Licensing" and "Experimenters Licensing".

For visitor's licences, the fee is EUR 12.00 euros (approx GBP £9.50) if the permit is required for one to three months and 12.00 euros for every subsequent three months renewed. The licence is free if it is for less than a month in a year.
Licensing authority: The Commission for Communications Regulation, Licensing Section, Abbey Court, Irish Life Centre, Lower Abbey Street, Dublin 1, Ireland; **website:** www.comreg.ie
Tel: +353 1 804 9600; **fax:** +353 1 804 9680.
E-mail: info@comreg.ie

ISRAEL (CEPT, HAREC)

Israel accepts the CEPT licence and the HAREC.

The CEPT licence is valid for up to three months in Israel, but if you have a CEPT licence and you are working in the country for a period of more than one year you may apply for a full Israeli callsign in the series 4Z8. Applications should be made to the Ministry of Communications and each application is decided on its own merit. You will need a letter from your employer, proof of temporary residency, proof of valid work permit issued by the Department of Labor and Welfare and a 'Temporary Resident' visa issued by the Ministry of the Interior.

For those who need to apply for a reciprocal licence, a "Questionnaire for Applicants for Reciprocal Amateur Radio License" is available on the Ministry of Communications website at www.moc.gov.il (click on "English" at the top right, then on "Forms and Procedures". A "Table of Frequencies and Conditions for Use by Radio Amateurs" is also available; click on "Frequencies").

Any queries regarding amateur radio licensing should be addressed to Ms Miriam Stessel, Radio Licensing Department, Spectrum Management Division at the Ministry of Communications.

The Israel Amateur Radio Club (IARC) is also authorised to issue short-term visitor's licences. Fill in 'Licensing form 006' which can be found on the IARC website by going to www.iarc.org/audible.html then clicking on "Licensing and CEPT". When completed, send the form to Israel Amateur Radio Club, PO Box 17600, Tel Aviv 61176, Israel.

Licensing authority: Ministry of Communications, Spectrum Management Division, PO Box 29107, Tel Aviv 61290, Israel;
website: www.moc.gov.il
Tel: +972 3 519 8270;
fax: +972 3 519 8103.
E-mail: stassel@moc.gov.il

ITALY (CEPT)
If you do not hold a CEPT Licence, you should contact the Italian Ministry of Communications direct to request a visitor's licence. You may ask the national amateur radio society, ARI, for support. Usually a reciprocal permit is granted, but it may take up to a few months. For help with the application

contact Associazione Radio-amatori Italiani (ARI), Via Domenico Scarlatti 31, I-20124 Milano, Italy; tel: +39 2 669 2192; fax: +39 2 6671 4809; e-mail: segreteria@ari.it Their website is at www.ari.it

Licensing authority: Ufficio Relazioni con il Pubblico (URP), Ministero delle Comunicazioni, Viale America, 201 - 00144 Roma, Italy;
website: www.comunicazioni.it
Tel: +39 6 5444 2100; **fax:** +39 6 5444 0014.
E-mail: urpcom@comunicazioni.it

IVORY COAST (COTE D'IVOIRE)
To apply for a licence you need to fill in two application forms in triplicate and provide copies of your home licence, your passport personal details page and visa and two passport-type photographs.

The application forms can be obtained from the IARU member society, the Association des Radio Amateurs Ivoiriens (ARAI), PO Box 2946, Abidjan 01, Côte d'Ivoire; tel: +225 21 243 346 or e-mail: tu2ci@arai-ci.org The Society's website is at

Italy, I, call districts.

www.qsl.net/tu2ci and the President is Jean-Jacques Niava, TU2OP, tel: +225 21 242 123 (home), +225 21 240 991 (work) or e-mail: tu2op@africamail.com

Members of the ARAI will assist you with the application procedure which may involve visits to the PTT, police and security offices. It is likely to take about one month before the licence is received, although it is possible you may be able to get on the air earlier, once the callsign has been allocated. The licence fee is CFA 25,000 francs (approx GBP £30) for a one-year licence.

Licensing authority: Agence des Télécommunications de Côte d'Ivoire (ATCI), Marcory Anoumanbo 18, BP 2203, Abidjan 18, Côte d'Ivoire; **website:** www.atci.ci
Tel: +225 20 3443 73 / 74 / +225 20 345 986; **fax:** +225 20 344 375.
E-mail: courrier@atci.ci

JAMAICA

It is relatively straightforward to obtain a visitor's licence in Jamaica. An 'Application for Alien Amateur Radio Station Permit' form ('Form C') is available on the Jamaica Amateur Radio Association (JARA) website at www.jamaicaham.org Complete and send the form to the Spectrum Management Authority (SMA).

You should enclose copies of both your passport and your home licence. If either is in a language other than English, it must be translated and the translation notarised. Apply two to three months before your visit in order to have time to receive your licence in the mail. If you are taking your own equipment into the country, you are advised to ask to have the equipment listed on the licence in order to help with possible customs inspections when entering the country.

The three-month visitor's licence is of the type 6Y/home call, but it is also possible to obtain a special event callsign (e.g. for contests). 6Y5 calls are reserved for Jamaican nationals, but any other format is possible (e.g. 6Y3Z).

The licence fee must be paid before the licence can be issued. Check the current licence fee with the SMA and send a certified bank cheque (check) made payable to "Spectrum Management Authority" with your application. If you wish to have a special event callsign you will need to pay for two licences.

The licensing officer is Mr Lloyd Matheson.
Licensing authority: Manager Administration, Spectrum Management Authority, 2nd Floor, VMBS Building, 53 Knutsford Boulevard, Kingston 5, Jamaica; **website:** www.sma.gov.jm
Tel: +876 929 8550 / 8520; **fax:** +876 960 8981.
E-mail: info@sma.gov.jm *or* lmatheson@sma.gov.jm

JAPAN

Japan has four classes of amateur radio licence, with power and frequency band restrictions that may seem unusual to Western radio amateurs. The Japan Amateur Radio League (JARL) English-language website explains these in great detail at www.jarl.or.jp/English/2_Outline/A-2-0.htm

Japan only has reciprocal licensing agreements with nine countries: Australia, Canada, Finland, France, Germany, Ireland, Korea, Peru and USA (but not the UK, for example). Note that those who hold a valid amateur radio licence in any of those countries are eligible to operate an amateur station in Japan, *regardless of their nationality*. British radio amateurs, therefore, may obtain a Japanese licence if they hold, for example, a current US or Irish licence.

Please refer to the JARL English-language website for full details of how to apply for a visitor's licence in Japan. What follows is a brief synopsis.

There are three possible licence classes for visitors. The first is a 50-watt portable / mobile licence that can be used anywhere in Japan. A separate 200-watt licence for a single *fixed* address is necessary if you wish to operate with more than 50 watts. You may hold both licences, but a fee is charged for each. These applications may be made via JARL. The third possibility is only suitable for those long-term residents of Japan who wish to use more than 200W. For this high-power licence, applications must be made using a Ministry of Internal Affairs and Communications (MIC) form filled out in Japanese. This form is sent to the regional Bureau of Telecommunications (BT) in the district in which you will operate. On-site inspection of your station by BT takes place and the whole procedure may take more than three months before the licence is granted.

For the 50W portable / mobile or 200W fixed station licences, you should apply via JARL, signing an authorisation form (JARL-04-09A) allowing JARL to submit the application on your behalf. This form, as well as the licence application form, are available on the JARL website. The licence fee(s) may be paid either by bank transfer to JARL's account or by International Money Order payable in yen to the Japan Amateur Radio League. The 50-watt portable / mobile licence is 8000 yen (approx GBP £40), the 200-watt fixed-only licence 11,800 yen (approx GBP £60). You must submit proof that the funds have been sent when sending in your application.

The licence(s) will be sent to the address given for the fixed station. However, if you are applying only for the portable / mobile licence, you may use JARL's address and pick up your licence at

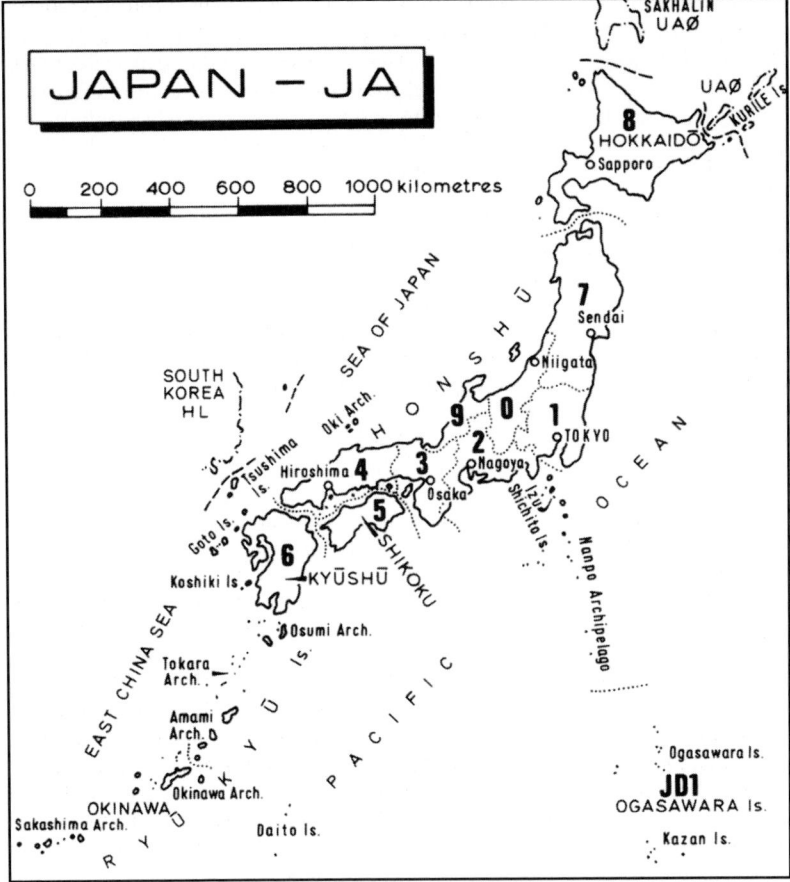

Japan, JA, call districts.

the JARL office in Tokyo by presenting your passport or other ID.

Submit your application, at least two months before the date the licence is required, to: Japan Amateur Radio League, International Section, 1-14-5, Sugamo, Toshima-ku, Tokyo 170-8073, Japan; tel: +81 3 5395 3106; fax: +81 3 3943 8282; e-mail: intl@jarl.or.jp

For further details please refer to www.jarl.or.jp/English/3_Application/A-3.htm, where you may also download the various application forms.

Licensing authority: Bureau of Telecommunications (BT), Ministry of Internal Affairs and Communications, 1-2, Kasumigaseki 2-Chome, Chiyoda-ku, 100-8926 Tokyo, Japan;
website: www.tele.soumu.go.jp
Tel: +81 3 5253 5923; **fax:** +81 3 5253 5925
E-mail: info@bt.soumu.go.jp

JORDAN

Visitor's licences, with JY9 two-letter callsigns, are issued by the Royal Jordanian Radio Amateurs Society (RJRAS).

Applicants should fill out the appropriate form at RJRAS headquarters and provide the following information: type of station licence required (fixed, portable or mobile), full name, home address, passport number, date and place of issue, nationality, date of birth, type of transmitter, frequency range and power, class of home licence, home callsign, location of operation in Jordan, photocopy of home licence indicating class, date of issue, length of validity, and period of stay in Jordan.

The operator's licence may take several weeks to be issued. Once it is in hand, you can then apply for a station licence.

The licence fee is JOD 60 Jordan dinars (approx GBP £43) for an annual licence valid from 1 January to 31 December.

Contact the RJRAS Secretary, Mohammad Balbeisi, JY4MB, c/o Royal Jordanian Radio Amateur Society, PO Box 2353, Amman 11181, Jordan; tel: +962 6 666235 or +962 6 810766; fax: +962 6 814566; e-mail: jy4mb@yahoo.com

Licensing authority: Telecommunications Regulatory Commission (TRC), PO Box 850967, Amman 11185, Jordan; **website:** www.trc.gov.jo
Tel: +962 6 586 2020; **fax:** +962 6 586 3641.
E-mail: trc@trc.gov.jo

KAZAKHSTAN

No specific information on amateur radio licensing is available, but the Head of the Licensing, Standardisation and Certification Division at the Agency for Informatization and Communication (AIC) is Ms Zulfiya Khudaiberghenova. Her address is Room 775, AIC, Block A, Ministries Building in Astana (see below).

The AIC website has a section in English for "Application Forms" under "Radio Frequency Spectrum", but this was 'Under Construction' at the time of compiling this book.

Licensing authority: Agency of the Republic of Kazakhstan for Informatization and Communication (AIC), Block A, Ministries Building, Left Bank,

Astana 010000, Kazakhstan;
website: www.aic.gov.kz
Tel: +7 7172 74 1015 / +7 7172 75 7709;
fax: +7 7172 74 1003 / +7 7172 24 0611.

KENYA

After many years when it was very difficult for visitors to obtain a Kenyan licence, it is now very much easier, largely thanks to strenuous efforts over a period of several years by members of the Amateur Radio Society of Kenya.

An "Application For Amateur Radio Station License" (Form RF2) is available for download from the Communications Commission of Kenya (CCK) website by clicking on "Application Forms" and then "Radiocommunication Application Forms".

You need to attach photocopies of your passport pages showing nationality, date of issue and expiry, name, photograph and Kenyan visa, and a certificate of a recognised Radio Amateur Examination or valid and current licence. Note that all photocopies *must* be certified by a Commissioner for Oaths. Uncertified documents will not be accepted and presentation of originals is no substitute as certified copies are required to be kept on file. Proficiency in Morse code is no longer required for any Kenyan licence.

The application fee is KES 1000 Kenyan shillings and the licence fee 2000 shillings (total approx GBP £24.50) and the licence is valid for a period of one year maximum.

Licensing authority: The Director General, Communications Commission of Kenya (CCK), Waiyaki Way (opposite Kianda School), PO Box 14448, Westlands 00800, Nairobi, Kenya;
website: www.cck.go.ke
Tel: +254 20 434 9111 / 424 2000;
fax: +254 20 445 1866 / 434 8204.
E-mail: info@cck.go.ke

KIRIBATI

An amateur radio licence application form can be found on the website of the Telecommunications Authority of Kiribati, TAK (Click on "Sections" under "Main Menu", at the top left of the home page, then on "Applications", and "Forms".

The Chief Licensing Officer at TAK is Mote Terukaio, T30MT.

Licensing authority: Telecommunications Authority of Kiribati, PO Box 252, Betio, Tarawa, Kiribati; **website:** www.tak.ki
Tel: +686 2532 or +686 25488; **fax:** +686 25432.
E-mail: enquiries@tak.ki

KOREA (DPRK, NORTH KOREA)

It is considered to be impossible to obtain an amateur radio licence in North Korea. The contact details of the licensing authority, as listed on the ITU website, are given below, for what it's worth!

Licensing authority: Ministry of Posts and Telecommunications, Oesong-dong, Central District, Pyongyang, Democratic People's Republic of Korea.
Tel: +850 2 381 3180; **fax:** +850 2 381 4418.
E-mail: mptird@co.chesin.com

KOREA (ROK, SOUTH KOREA)

Two slightly different application forms (depending on whether you plan to use your own transceiver or an existing station in Korea) are available for downloading from the Korean Amateur Radio League (KARL) website. Click on "English" at the top right of the home page, then "Operation in Korea".

If you intend to use your own transceiver, inspection of the station is required. No inspection is required if you intend to use an existing Korean station (but you must give the callsign of that station in your application).

The completed application form should be returned to the Korean Amateur Radio League (*street address:* KARL Building, 3F, Yang Jae-Dong 267-5, Seocho-Gu, Seoul 137-130, Korea; *postal address:* CPO Box 162, Seoul 100-601, Korea); tel: +82 2 575 9580; fax: +82 2 576 8574; e-mail: karl@karl.or.kr at least 60 days before the start of your visit. Also required are a copy of your home licence, a copy of your passport page that includes your photograph, a table showing the permissible maximum output power by class in your home country, and the licence fee (KW 80,000 for under 50W or KW 100,000 for above 50W). The licence fee should be paid to account number 204-22-05221-7 at the Korea Exchange Bank, Poi Branch.

Licensing authority: Korea Communications Commission, 100 Sejongro, Chongro-gu, Seoul 110-777, Republic of Korea; **website:** www.kcc.go.kr
Tel: +82 2 750 1710; **fax:** +82 2 750 1729
E-mail: parksk@mic.go.kr

KOSOVO

Kosovo (Kosova) declared independence from Serbia on 17 February 2008, and its independence was recognised by a number of major countries, including the USA, UK and several other EU member states. At the time of going to press, however, Kosovo was not a member of the UN or the ITU and did not have an ITU-allocated radio communications prefix. Radio amateurs in Kosovo were continuing to use the YU8 prefix previously used when Kosovo was a province of Serbia. However, after Kosovo's declaration of independence, Serbia started issuing YU8 callsigns to radio amateurs in other parts of Serbia.

Details of the radio spectrum regulatory body,

the Autoriteti Rregullator i Telekomunikacionit (ART), or Telecommunications Regulatory Authority of Kosovo, are given below, but there is no information about amateur radio in Kosovo on their English-language website.

Licensing authority: Autoriteti Rregullator i Telekomunikacionit (ART), St Pashko Vasa, N.12, 10000 Prishtina, Republic of Kosovo;

website: www.art-ks.org

Tel: +38 212 345.

E-mail: info@art-ks.org

KUWAIT

In order for visitors to obtain a Kuwaiti licence, they must first hold a USA FCC amateur licence. An application form is available from the Kuwait Amateur Radio Society (KARS) headquarters, PO Box 5240, Safat 13053, Kuwait; tel: +965 533 3761 / 3762; fax: +965 531 1188; e-mail: 9k2ra@kars.org (for those in Kuwait KARS HQ is located at Surra area, District 2, Street 1, Substreet 12, Building 2).

After completion, the application form should be submitted to the KARS manager along with a copy of your passport showing the visa (for those on a short visit), or a copy of your Civil ID Card for those residing in Kuwait. You should also provide the specifications and serial numbers of your equipment. KARS refers the application to the Ministry of Communication and the process may take from one to 10 days. The licence fee is KWD 5 dinars (approx GBP £9.50) per year, payable to the Ministry of Communications.

Licensing authority: Ministry of Communications, PO Box 318, 11111 Safat, Kuwait;

website: www.mockw.net

Tel: +965 481 9033; **fax:** +965 484 7058.

E-mail: moc@mockw.net

KYRGYZSTAN

Visitor's licences are handled by the Amateur Radio Union of Kyrgyzstan (ARUK), PO Box 745, Bishkek 720017, Kyrgyzstan (President Ivan Udovin, EX2A). Copies of passport, home licence, two photographs, and address in Kyrgyzstan from where you will be operating are required.

Serge Guzev, EX0M, PO Box 444, Bishkek 720075, Kyrgyzstan (e-mail: ex0m@mail.ru) and George Lazarev, EX2M, PO Box 2177, Bishkek 720021, Kyrgyzstan (e-mail: ex2m@mail.ru) have agreed to help visitors to Kyrgyzstan with their licence applications.

Licensing authority: National Information Resources Technology and Communications Agency, 7B Baytik Baatyr Street, Bishkek 720005, Kyrgyzstan.

Tel: +996 312 544 103; **fax:** +996 312 544 105.

E-mail: nta@infotel.kg

LAOS

Foreigners holding a residence or work permit in Laos are usually able to obtain an amateur radio licence but it is more difficult for short-term visitors. The stumbling block is that it is necessary to have a guarantee letter from a Laos-registered state-owned enterprise to act as a local sponsor, and this is normally only available to those with work permits. If you wish to try, the following is believed to be the most up-to-date information.

Apply in writing, giving the frequencies and modes required (it is suggested you request "all Amateur Service frequencies allocated in ITU Region 3" and "all modes"), and listing the make and model of transceiver and type of antenna(s) to be used, the place of operation and the dates the licence is required. The maximum power granted is 100 watts, so do not list a linear amplifier among the equipment. You must also give your name, home address, nationality, passport number, class of home licence and callsign, as well as photocopies of your passport and home licence and the all-important local sponsor's letter of guarantee.

Send the completed application to both the Ministry of Post and Telecommunications (details below) *and* the KPL (Khaosan Pathet Lao) News Agency, 80 Setthathirath Road, PO Box 3770, Vientiane, Lao PDR; tel: +856 21 215 402, 212 447 *or* 251 090; fax: +856 21 212 446; e-mail: kplcab@laonet.net KPL has a website at www.kpl.net.la

The licence fee is USD $150 for a three-month licence, making this one of the more expensive amateur licences available, and the application is likely to take around three weeks.

Licensing authority: Head, Frequency Manager Division, Department of Post and Telecommunications, Jawaharal Nehru Street, 0100 Vientiane, PDR Laos.

Tel: +856 21 412 299; **fax:** +856 21 412 279.

E-mail: laofreqm@laotel.com

LATVIA (CEPT)

If you do not have a CEPT Licence, you may apply for a temporary visitor's licence. Request an application form from the SPRK and, when completed, return the form with a passport-size photo and copy of your home licence. The licence fee is approximately USD $10.

Licensing authority: Public Utilities Regulatory Commission (SPRK), Brivibas Street 55, Riga LV-1010, Latvia; **website:** www.sprk.gov.lv

Tel: +371 6709 7200; **fax:** +371 6709 7277.

E-mail: sprk@sprk.gov.lv

LEBANON

Lebanon has no reciprocal licensing agreements in place and so requests for a visitor's licence are dealt with on a case-by-case basis. The Radio Amateurs of Lebanon (RAL) website at http://ral.org.lb has a section on licensing from where it is possible to download licence application forms (in Arabic).

Alternatively, foreign residents in Lebanon may take the local radio amateur examinations, which are offered in English and French. See the RAL website for full details.
Licensing authority: Ministry of Telecommunications Building, 1st Floor, Riad El Solh (Banks) Square, Beirut Central District, Beirut, Lebanon; **website:** www.mpt.gov.lb
Tel: +961 1 979 137 / 161;
fax: +961 1 979 120 / 164.
E-mail: spectre@mpt.gov.lb

LESOTHO
Visitors' licences are issued in Lesotho with little formality. The application fee plus licence fee is ZAR 250+60 South African rand (R310, approx GBP £21). For visitors' licences, the callsign granted is 7P8/own call.
Licensing authority: Mr Mamofolo Kobeli, Lesotho Communications Authority (LCA), Moposo House, 6th Floor, Kingsway Road, PO Box 15896, Maseru 100, Lesotho.
Tel: +266 22 326 784 / +266 22 325 595;
fax: +266 22 310 984.
E-mail: mkobeli:lta.org.ls *or* admin@lta.org.ls

LIBERIA
It is possible to receive a visitor's licence in Liberia. You should apply by letter at least two months before the licence is required, with the usual documentation as well as a letter attesting to your good character (e.g. from the police, your national amateur radio society or your local church.) If you know a local resident of Liberia who can act as your sponsor or agent, so much the better.
Licensing authority: Radio Regulatory and Licensing Bureau, Ministry of Posts and Telecommunications, GPO Central Complex, Monrovia, Liberia.
Tel: +231 221 830; **fax:** +231 227 838.

LIBYA
A small number of licences has been issued in Libya in recent years, most notably 5A7A for a German-led multi-national operation from the Janzour Tourist Village, west of Tripoli, in 2006. It would probably be best to have an Arabic-speaking radio amateur make any application.
Licensing authority: Spectrum Management and Licensing Directorate, PO Box 81686, Tripoli, Libyan Arab Jamahiriya.
Tel: +218 21 360 4101; **fax:** +218 21 360 4102.

LIECHTENSTEIN (CEPT, HAREC)
Liechtenstein accepts the CEPT Licence and the HAREC. (An application form available for download from the website of the Liechtenstein Office for Communications at www.ak.llv.li makes it clear that the HAREC is accepted, even though Liechtenstein is not among a list of countries, dated 30 April 2008, that have implemented CEPT Recommendation T/R 61-02 on the ERO website.)

To download the (German language) application form, click on "Amateurfunk, Antrag" under "Onlineschalter - AK" at the bottom right of the home page.
Licensing authority: Office for Communications, Kirchstrasse 10, Post Box 684, 9490 Vaduz, Principality of Liechtenstein; **website:** www.ak.llv.li
Tel: +423 236 6488; **fax:** +423 236 6489.
E-mail: office@ak.llv.li

LITHUANIA (CEPT)
Lithuanian licenses are issued by RRT (Rysiu Reguliavimo Tarnyba, the Communications Regulatory Authority). The RRT website has information about licensing and an application form in English. Go to www.rrt.lt, click on the British Union Jack flag if the website is opened in Lithuanian, then click on "For radio amateurs" in the "For consumers" section.
Licensing authority: Communications Regulatory Authority (RRT), Algirdo Street 27, LT-03219 Vilnius, Lithuania; **website:** www.rrt.lt
Tel: +370 5 210 5633 / +370 5 216 1177;
fax: +370 5 216 1564.
E-mail: rrt@rrt.lt

LUXEMBOURG (CEPT)
Luxembourg accepts the CEPT Licence. If you do not have one, apply via the Institut Luxembourgeois de Régulation (ILR). There is an application form

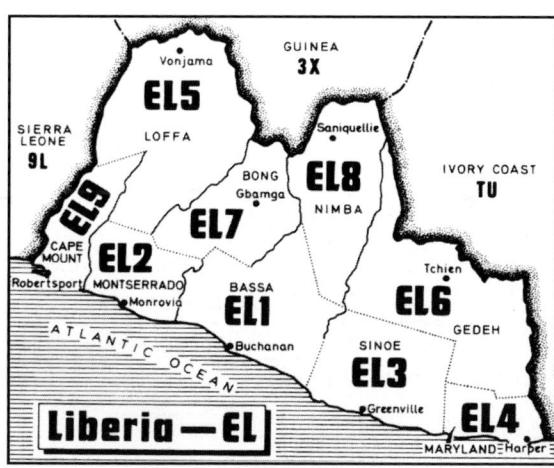
Liberia, EL, call districts.

(in French) available for downloading from the ILR website. Click on "Fréquences", then "Radio-Amateurs" for a list of several forms and brochures of interest to radio amateurs.

Licensing authority: Institut Luxembourgeois de Régulation (ILR), 45 Allée Scheffer, 2922 Luxembourg, Luxembourg; **website:** www.ilr.lu
Tel: +352 4588 451 / +352 4588 4521;
fax: +352 45 88 45 88.
E-mail: info@ilr.lu

MACAU

If you intend to take your own equipment into Macau, it is necessary to have a station inspection before the licence is granted. Apply at the Bureau of Telecommunications Regulation (DSRT) on Avenida da Praia Grande, complete the application form, pay the licence fee, and arrange a convenient time for the station inspection.

For assistance in the process, contact the IARU member society, Associacao dos Radioamadores de Macau [ARM], Box 6018, Macau SAR, China; tel: +853 8688 1515; fax: +853 8880 120; e-mail: arm@arm.org.mo

Licensing authority: Bureau of Telecommunications Regulation (DSRT), Avenida da Praia Grande 789-795, 3rd Floor, Macao, Macao SAR, China;
website: www.dsrt.gov.mo
Tel: +853 8396 9183 (Monday to Friday, 0900-1300 & 1430-1745); **fax:** +853 8396 9166.
E-mail: ifx@dsrt.gov.mo

MACEDONIA (FORMER YUGOSLAV REPUBLIC OF MACEDONIA, FYROM) (CEPT, HAREC)

If you are using the CEPT Licence you should use the prefix Z38/ before your own callsign. Macedonia also accepts the HAREC and you may apply for a Z38 one- or two-letter visitor's call using the application form that is available for download from the Macedonian Agency for Electronic Communications (AEC) website at www.aec.mk Click on the UK flag for English, then "Legislation", "Radio frequency spectrum" then "07. Application form on Issuing an Authorisation for Radio Frequency Utilisation in Radio Amateur Service".

Licensing authority: Agency for Electronic Communications, 13 Dimitrie Cupovski Street, 1000 Skopje, The Former Yugoslav Rep. of Macedonia;
website: www.aec.mk
Tel: +389 2 328 9200; **fax:** +389 2 322 4611.
E-mail: contact@aec.mk

MADAGASCAR

A visitor's licence is available in Madagascar from OMERT, the Office Malagasy d'Etudes et de Régulation des Télécommunications. To apply, you should write in French requesting a licence, enclosing three identity photographs, a copy of your home licence, a copy of your passport and Madagascar visa and, if you intend to take your own equipment into the country, a list of the manufacturer, model and serial numbers of each piece of equipment and a technical description of all transmitting equipment (e.g. copied from the handbook).

If you stay at 'La Villa Suede' (see the Rental Stations section of this book) Åke Rosvall, 5R8FU, can help you to apply for a 5R8 licence. Send Åke copies of all the above along with a payment of approximately USD $100. The licence will be issued together with a document for the import and export of a rig (if necessary).

Licensing authority: Office Malagasy d'Etudes et de Régulation des Télécommunications (OMERT), Route des Hydrocarbures Alarobia, BP 99991, MPT 101 Antananarivo, Madagascar;
website: www.omert.mg
Tel: + 261 20 224 2119; **fax:** + 261 20 232 1516.
E-mail: omert@moov.mg

MALAWI

It is easy to get a visitor's licence in Malawi and MACRA, the licensing authority, is one of the most efficient in Africa. Applications are accepted by fax and the licence fee is MWK 1000 kwachas

MACRA, the Malawi licensing authority, is one of the most efficient in Africa.

(approx GBP £3.50) for a one year licence.
Licensing authority: The Director General, Malawi Communications Regulatory Authority (MACRA), MACRA House, Salmon Amour Road, Private Bag 261, Blantyre, Malawi;
website: www.macra.org.mw
Tel: +265 1 623 611; **fax:** +265 1 623 890.
E-mail: dg-macra@malawi.net

MALAYSIA

Malaysia has reciprocal licensing agreements with 13 countries: Thailand, Laos, Indonesia, Vietnam, Myanmar, India, China, Mongolia, United Kingdom, Germany, Switzerland, Finland and Czechoslovakia [sic]. For those countries where there is no reciprocal agreement, applications are dealt with on a case by case basis. In practice, amateurs from most major countries will be granted a licence without any problem.

In Malaysia, the amateur radio licence is known as an 'Amateur Station Apparatus Assignment' ('ASAA', or simply 'AA'). Applications should be sent to the Malaysian Communications and Multimedia Commission (MCMC) office covering the location from which you intend to operate (see below). To apply for a licence, you must complete an MCMC application form and attach a sketch of the aerials to be used, documentary proof of citizenship (copy of passport), copy of your home licence or Radio Amateur's certificate, a letter of reference from MARTS (the Malaysian Amateur Radio Transmitters' Society) or - if applying in Sabah or Sarawak - a letter of reference from two Malaysian Class A amateurs, and finally a

PHOTO: JA2KLT

Steve Telenius-Lowe, 9M6DXX (left), and John Plenderleith, 9M6XRO (right), discuss licensing issues with MCMC Assistant Director for Sabah & FT Labuan Region, Mr Amid Drajak.

Statutory Declaration of Secrecy, signed before a Commissioner of Oaths, Magistrate or Justice of the Peace.

A 55-page pdf document, *Guidelines for Amateur Radio in Malaysia*, may be downloaded from the licensing authority's website at www.skmm.gov.my Click on "Codes & Guidelines" under "Quick Links" on the home page, then scroll down the page to locate the file. Samples of the application form, letter of reference and Declaration of Secrecy are given as appendices to *Guidelines for Amateur Radio in Malaysia*, but blank forms are no longer available on the website. A blank application form and Declaration of Secrecy form may instead be downloaded from the Hillview Gardens Amateur Radio Club website at www.qsl.net/9m6aac/aaclic.htm

Visitor's licences are issued for a period of three months or one year as requested, and the callsign allocated is 9M2/own call if you apply in Peninsula Malaysia, 9M6/own call if you apply in the state of Sabah, or 9M8/own call if you apply in the state of Sarawak (only residents of Malaysia are allocated 'full' callsigns). All three prefixes may be used anywhere in Malaysia: the second digit in the callsign only indicates

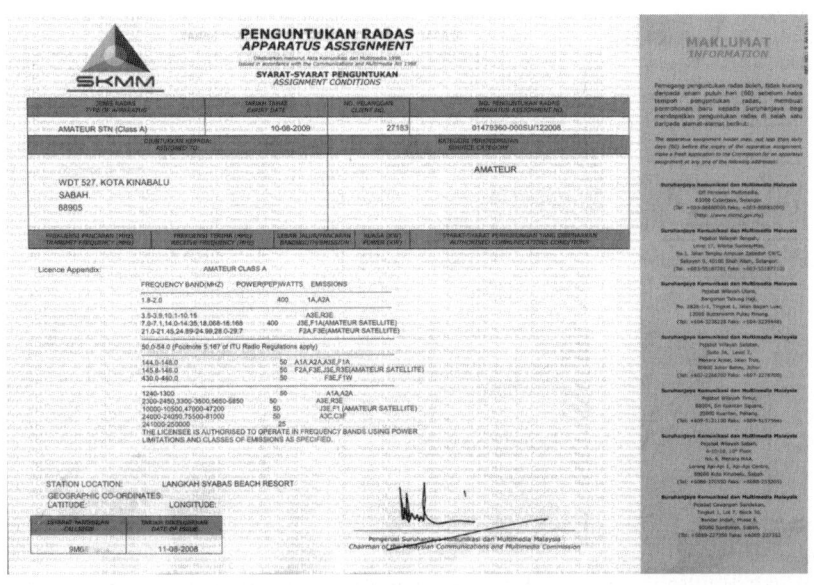

The Malaysian licence.

where the licence was issued. The licence fee is made up of an application fee of MYR 60.00 Malaysian ringgit, plus a monthly fee of MYR 3.00, i.e. a total of MYR 69.00 (approx GBP £11.25) for a three-month licence, or MYR 96.00 (approx £15.60) for a one-year licence.

If you have already completed all the documentation, a licence can be issued to you while you wait, or within one or two working days, particularly in the regional offices.

Note about operating from Pulau Layang Layang (Malaysian Spratly Islands): Although the Malaysian licence is valid for Layang Layang, it is *essential* also to obtain *written* permission from the Malaysian National Security Agency *before* arriving on the island. Failure to do so will result in operating permission being denied by the manager of the island resort and by the commander of the navy base on the island.

Licensing authority: Suruhanjaya Komunikasi dan Multimedia Malaysia (Malaysian Communications and Multimedia Commission, MCMC); **website:** www.skmm.gov.my
Head office (9M2): MCMC, 63000 Cyberjaya, Selangor, Malaysia;
Tel: +60 3 8688 8000;
fax: +60 3 8688 1000 / +60 3 8688 1001.
Penang (9M2): MCMC, Unit 3 Level 11, Menara UMNO, 128 Jalan Macalister, 10400 Pulau Pinang, Malaysia;
Tel: +60 4 227 1657; **fax:** +60 4 227 1650.
Johor Bahru (9M2): MCMC, Suite 7A Level 7, Menara Ansar, Jalan Trus, 80000 Johor Bahru, Johor, Malaysia;
Tel: +60 7 226 6700; **fax:** +60 7 227 8700.
Kuantan (9M2): MCMC, B8004 Tingkat 1, Sri Kuantan Square, Jalan Telok Sisek, 25200 Kuantan, Pahang;
Tel: +60 9 515 0078; **fax:** +60 9 515 7566.
Sabah (9M6): MCMC, 6-10-10, 10th floor, No 6, Menara MAA, Lorong Api-Api 1, Api-Api Centre, 88000 Kota Kinabalu, Sabah, Malaysia;
Tel: +60 88 270 550; **fax:** +60 88 253 205.
Sarawak (9M8): MCMC, Level 5 (North), Wisma STA, 26 Jalan Datuk Abang Abdul Rahim, 93450 Kuching, Sarawak;
Tel: +60 82 331 900; **fax:** +60 82 331 901.

MALDIVES

It is relatively easy to get an amateur radio licence in the Maldives. The website of TAM, the Telecommunications Authority of Maldives, has detailed information: click on "Services", then "Amatuer License" [sic].

Request an application form from TAM. Complete the form and send it to TAM (e.g. by fax) with a copy of your home licence. You must state in the application the island from which you will

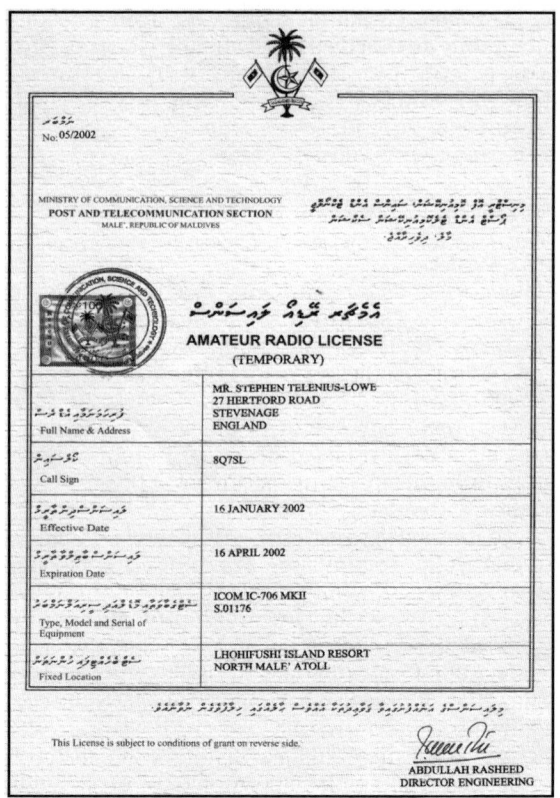

The Maldives licence.

be operating.

There are two types of licence, a Temporary Licence (valid for three months), and a Yearly Licence. The licence fee is MVR 125 Maldives rufiyaa (approx GBP £5) for the Temporary Licence or MVR 200 rufiyaa (approx GBP £7.85) for the Yearly Licence. The licence will be issued within three working days of receipt of the licence fee.

If you apply in sufficient time, TAM will arrange for your licence to be sent to the local tour operator who should be able to deliver it to the island on which you are staying.

Licensing authority: Telecommunications Authority of Maldives, Telecom Building, Husnuheena Magu, Male', Maldives;
website: www.tam.gov.mv
Tel: +960 332 3344; **fax:** +960 332 0000.
E-mail: secretariat@tam.gov.mv

MALI

It is possible to obtain a visitor's licence in Mali but it takes some time and effort. Write an application letter in French with copies of your home licence, passport, Mali visa, and technical details of your equipment and antenna(s), e.g. photocopy of technical specifications from the manual. All correspondence should be in French.

Send two copies of everything to Sotelma. The IARU member society may be able to help: contact Club des Radioamateurs et Affiliés du Mali (CRAM), PO Box 9A, Kati, Mali; tel / fax: +223 227 2114; e-mail: tz6hy@caramail.com (President of CRAM Hamadoun Yattara, TZ6HY).

Licensing authority: Societé de Télécommunication du Mali (Sotelma), Direction Commerciale (Agence Centre commercial), Rue Karamoko Diaby Centre Commercial, BP 740, Bamako, Mali; **website:** www.sotelma.ml **Tel:** +223 222 7738; **fax:** +223 222 9845. **E-mail:** info@sotelma.ml

MALTA (HAREC)

For well over a decade Malta has apparently been on the point of accepting the CEPT Licence, but as of mid 2008 this has still not happened and therefore it is still necessary to apply for a visitor's licence. When Malta will finally accept the CEPT Licence is anybody's guess: the present situation has been going on for so long that CEPT has even added a footnote to Appendix II of T/R 61-01, which reads, "A revision to the current legislation is still being considered by the Malta Administration. Hence, for the time being, visitors still have to apply for a licence and callsign."

In practice, Maltese visitor's licences (with 9H3 callsigns) are not difficult to obtain and Malta *has* accepted the HAREC, so it is possible to obtain a full licence for those staying in Malta for three months or making repeated visits there.

The Malta Communications Authority (MCA) website has an application form available for downloading at www.mca.org.mt/infocentre/openarticle.asp?id=733&pref=31

Complete the form and send it with a photocopy of your current home licence, details of your date of arrival in and departure from Malta and copy of passport. The licence fee is EUR 13.90 euros (approx GBP £11) and the licence will be issued quickly after receipt of the fee.

Note that although the headquarters of MCA is at Pinto Wharf on Valletta Waterfront, amateur radio licensing is still handled from the old Prime Minister's Office building in Merchants Street.

Licensing authority: Director of Corporate Services, Malta Communications Authority, Evans Building, Merchants Street, Valletta, Malta GC; **website:** www.mca.org.mt **Tel:** +356 21 247 224 – 228; **fax:** +356 21 247 229. **E-mail:** info@mca.org.mt

MARSHALL ISLANDS

Amateur radio licences are issued quickly, efficiently and free of charge to visitors. The Director of Communications at the Ministry of Transport and Communications in 2008 is Mr Carthney Laukon.

Licensing authority: Director of Communications, Ministry of Transport and Communications, PO Box 1079, Uliga Street, Majuro, MH 96960, Marshall Islands. **Tel:** +692 625 5010; **fax:** +692 625 5011. **E-mail:** freqman@ntamar.net

MAURITANIA

No current amateur radio licensing information available. Contact the Autorité de Régulation (ARE) in Nouakchott for details.

Licensing authority: Autorité de Régulation (ARE), Ilot Z Lot No 14, BP 4908, Nouakchott, Republique Islamique de Mauritanie; **website:** www.are.mr **Tel:** +222 529 1270; **fax:** +222 529 1279. **E-mail:** info@are.mr

MAURITIUS

Amateur radio licensing for visitors to Mauritius is now much easier than it was just a few years ago, although it can still be difficult to obtain a visitor's licence for any of the Mauritius dependencies, i.e. Rodrigues, Agalega and St Brandon.

Applications are made to the Information & Communication Technologies Authority (ICTA) but have to be approved by the security service and the government before a licence can be issued. All this takes time and you are therefore advised to apply for a licence at least three or four months before the date it is required, preferably six months.

To apply for a licence, fill in the official ICTA application form and attach a photocopy of your passport and licence, and two passport-size photographs. The application form can be found in pdf format on the ICTA website under "Radiocommunications" and "Application Forms". (A sample licence, with the full terms, conditions and schedule, can also be found on the website under "Specimen licences".) Send a copy of all the documents to the Secretary of the Mauritius Amateur Radio Society (MARS), c/o Mr Seewoosankar Mandary, 3B8CF, PO Box 104, Quatre Bornes, Mauritius. This will enable MARS to provide information if requested either by ICTA or the security personnel and help MARS to follow the progress of the application on your behalf. Note that the name and address of a contact person in Mauritius must be provided on the application form: MARS officials may be able to help with this requirement if you do not have friends or family on the island.

The visitor's licence is valid for one year and the licence fee is MUR 1000 rupees (approx GBP

£18). However, you should not send the fee to Mauritius at the time of application. If all goes well, you should be able to pick up your licence in person at the ICTA office in Port Louis and pay the licence fee then.

If you plan to operate from Rodrigues, or particularly Agalega or St Brandon, you are advised to apply for the licence in person at the ICTA office. Note that landing permits and other documents are necessary to visit Agalega and St Brandon.

You should not take any transmitting equipment into Mauritius until you have received approval from ICTA.

Licensing authority: Information & Communication Technologies Authority (ICTA), Level 12, The Celicourt, 6 Sir Celicourt Antelme Street, Port Louis, Mauritius; **website:** www.icta.mu
Tel: +230 211 5333/4; **fax:** +230 211 9444.
E-mail: icta@intnet.mu

MEXICO

The fact that you rarely hear US amateurs operating from Mexico, despite the proximity of the two countries, gives a clue that it is not that easy to obtain a visitor's licence in Mexico!

In fact, licences *are* issued, but it requires substantial paperwork and patience. There are different application forms to complete depending on whether you have a US licence or another foreign licence, and the application forms must be typewritten, not handwritten. There are links to the two application forms on the COFETEL website at www.cft.gob.mx/cofetel/html/agitec/requi/permiso.shtml, but neither form was actually available for downloading when checked between April and June 2008.

It is obligatory to have an invitation letter from a Mexican radio amateur, and anyway without the help of a Mexican amateur the application process would be much more difficult.

The procedure is long-winded but is described in detail in the web-log of Christian Buenger, DL6KAC, at http://ham-blog.de/reciprocal-licence-xe-permit.html

Licensing authority: Comisión Federal de Telecomunicaciones (COFETEL), Bosque de Radiatas No 42 - 4 piso, Bosques de las Lomas, Cuajimalpa, 05120 México DF, Mexico;
website: www.cft.gob.mx
Tel: +52 55 5015 4474 / 4485 / 4486;
fax: +52 55 5015 4047.
E-mail: contactocft@cft.gob.mx

MICRONESIA

Amateur radio licences for Pohnpei, Yap, Chuuk (Truk), and Kosrae are issued without difficulty from the FSM capital in Pohnpei. Apply to the Frequency Manager at the Department of Transportation, Communication and Infrastructure for an application form and submit it with a copy of your passport and home licence.

Licensing authority: Frequency Manager, Division of Communications, Department of Transportation, Communication and Infrastructure, Government of the Federated States of Micronesia, PO Box PS-2, Palikir, Pohnpei, FM 96941, Federated States of Micronesia;
website: www.fsmgov.org
Tel: +691 320 2865; **fax:** +691 320 5853.

MOLDOVA

It is possible to obtain a visitor's licence in Moldova. For assistance contact the IARU society, Asociatia Radioamatorilor din Moldova (ARM), PO Box 1414, MD 2043 Chisinau, Moldova; e-mail: arm@telco.md

Licensing authority: National Regulatory Agency in Telecommunications and Informatics, 4th Floor, 134 Stefan cel Mare Boulevard, MD 2012 Chisinau, Moldova; **website:** www.anrti.md
Tel: +373 22 251 312 / 315;
fax: +373 22 222 885.
E-mail: office@anrti.md

MONACO (CEPT)

Although Monaco has adopted CEPT Recommendation T/R 61-01 and therefore accepts the CEPT Licence, you *must* inform the Director of Telecommunications of all operation on the territory of Monaco under CEPT.

If you do not hold a CEPT Licence you may apply for a visitor's licence by contacting the same department.

From February 2007 the Director of Telecommunications is M Frederic Rue.

Licensing authority: Directeur des Télécommunications, Direction du Contrôle des Concessions et des Télécommunications, 23 Avenue Albert II, MC 98000 Monaco Cédex, Monaco; **website:** www.gouv.mc (click on "English", "The government", "Department of Facilities, Urban Planning and Environment", and finally [in right-hand panel] "Concession and Telecommunications Control Direction".)
Tel: +377 97 9 85 656 / +377 98 988 800;
fax: +377 97 985 657.
E-mail: frue@gouv.mc

MONGOLIA

Visitors to Mongolia may obtain a licence to operate Mongolian club or individual stations. Apply to the Mongolian Amateur Radio Society (MARS), PO Box 830, Ulaanbaatar 24, Mongolia; tel: +976 1 320 058; fax: +976 11 315 434; e-mail: jt1kaa@gmail.com

Please enclose an SAE and return postage for receipt of the application form. Complete the form and attach a photocopy of your home licence and an identity photograph. If you intend to bring your own equipment, you should provide the details to MARS with your application.

Licensing authority: Communications Regulatory Commission of Mongolia, Amarsanaa Street 26, Ulaanbaatar 210524, Mongolia; **website:** www.crc.gov.mn (in Mongolian Cyrillic characters only) **Tel:** +976 11 300 205 1237; **fax:** +976 11 327 720. **E-mail:** crc@mongol.net

MONTENEGRO

No specific details available about amateur radio licensing, although Ranko Boca, 4O3A, has hosted several foreign radio amateurs at his 'super station' (see 'Rental Stations' section of this book) and they have signed 4O/own call. Contact Ranko by e-mail: yt6a@cg.yu for further details.

The ITU lists the Ministry of Transport, Maritime Affairs and Telecommunications contact details as below.

Licensing authority: Ministry of Transport, Maritime Affairs and Telecommunications, Rimski Trg 46, 81000 Podgorica, Montenegro. **Tel:** +38 281 234 179; **fax:** +38 281 234 342.

MONTSERRAT

Amateur radio licensing is easy on Montserrat. These days visiting amateurs are issued with their own three-letter callsign, with your choice of two letters following the VP2M prefix, if available. All bands are permitted at a maximum power level of 1000W. Apply in person to the Ministry of Communications and Works. The licence fee is XCD EC$15 Eastern Caribbean dollars (approx GBP £2.85).

If you are renting his villa on Montserrat, Graham Dawes, VP2MDD / M0AEP, offers to help with the licensing procedure (see Rental Stations section for further details).

Licensing authority: Ministry of Communications and Works, Woodlands, Montserrat, BWI; **website:** www.gov.ms/commsworks **Tel:** +1 664 491 2521 / 2522; **fax:** +1 664 491 3475 / 6659. **E-mail:** mcw@gov.ms

MOROCCO

To apply for a visitor's licence contact the secretary of the Moroccan national society, Association Royale des Radio-Amateurs du Maroc [ARRAM], 12 Rue Ahmed Arabi, BP 299, Agdal, Rabat, Morocco; tel: +212 37 673 703; fax: +212 37 674 757. You will need several copies of your home licence, passport and identity photographs, as well as a list of manufacturer, model number and serial number of all equipment you will be importing.

Licensing authority: Agence Nationale de Réglementation des Télécommunications (ANRT), Centre d'Affaires, Boulevard Ar-Ryad, Hay Ryad, BP 2939, Rabat 10100, Morocco; **website:** www.anrt.ma **Tel:** +212 37 718 400; **fax:** +212 37 203 862.

MOZAMBIQUE

A visitor's licence is issued in Mozambique with little difficulty these days. The licence fee is USD $100. A specific callsign suffix may be requested; give two or three possible alternatives. As of mid-2008 the licensing officer was Mr Anselmo Ferrao.

Licensing authority: Instituto Nacional das Comunicações de Moçambique (INCM), Avenida Eduardo Mondlane 123/127, Caixa Postal 848, Maputo, Mozambique; **website:** www.incm.gov.mz **Tel:** +258 21 490 130 - 131 / +258 21 227 100; **fax:** +258 21 494 435. **E-mail:** ansferrao@ incm.gov.mz *or* info@incm.gov.mz

MYANMAR (BURMA)

It used to be impossible to get a visitor's licence in Myanmar; it is now merely very difficult.

According to the Posts and Telecommunications Department (PTD) website, its Radio Division "deals with ITU-R related activities such as

This Montserrat licence looks like it might date from the 1940s but in fact was issued in 1984.

radio frequency spectrum management and Licensing matters. The main duties are [among others] Issuing wireless station licenses and collecting fees [and] Prescribing radio frequency and supervising its usage."

The PTD website admits to the existence of amateur station licences in its table of licence fees, but does not state how much the fee is, nor how one can be obtained.

PTD has moved away from Yangon (Rangoon) and is now in the new capital of Nay Pyi Taw.

Licensing authority: Radio Division, Posts and Telecommunications Department (PTD), Ministry of Communications, Posts and Telegraphs, Building Number 2, Special Development Zone, Nay Pyi Taw, Myanmar;
website: www.mcpt.gov.mm/ptd/index.htm
Tel: +95 67 407 225; **fax:** +95 67 407 216
E-mail: dg.ptd@mptmail.net.mm

NAMIBIA
Amateur radio licensing is handled by the Namibian Communications Commission (NCC). An application form for a visitor's licence can be found easily on the NCC website: the completed form should be sent to the address below at least four or five weeks before your visit, with a *certified* copy of your current home licence and the licence fee of NAD $50.00 Namibia dollars (approx GBP £3.25). NCC's bank details, for payment of the licence fee, are given on the application form.
Licensing authority: Namibian Communications Commission (NCC), Ministry of Information and Broadcasting; **website:** www.ncc.org.na
Postal address: Private Bag 13309, Windhoek, Namibia.
Street address: Communication House, 56 Robert Mugabe Avenue, Windhoek.
Tel: +264 61 222 666; **fax:** +264 61 222 790.
E-mail: info@ncc.org.na

NAURU
Amateur radio licences are issued if you apply in person in Nauru. Take your original licence, a copy of the licence, and (if your country issues one) also a copy of your Radio Amateurs' Certificate.
Licensing authority: Ministry of Telecommunications, Nauru, Nauru;
website: http://naurugov.nr (contact details only)
Tel: +674 444 3133; **fax:** +674 444 3881.
E-mail:
minister.telecommunications@naurugov.nr *or* secretary.telecommunications@naurugov.nr

NEPAL
An application form for an amateur radio licence is available for downloading from the Ministry of Information and Communications website. Go to "License Application Forms" and scroll down to the bottom of the page.

After the form has been completed, you should take it in person to the Ministry of Information and Communications with your *original* licence, a copy of the licence, list of equipment and serial numbers, copy of passport and visa, and the licence fee. The licence is for 100W and is valid for the period of your visa.

The licence fee is calculated on *pairs* of frequencies, and is NPR 8400 Nepal rupees (approx GBP £66) for one pair of frequencies (bands), so for four bands you would pay 16,800 rupees (approx £132) or for all bands 10 to 80m 33,600 rupees (approx £264), making the Nepalese licence one of the more expensive available.
Licensing authority: Ministry of Information and Communications, Singh Durbar, Kathmandu, Nepal; **website:** www.moic.gov.np
Tel: +977 1 422 1647 / 5556 / 7515 / 7728; **fax:** +977 1 422 1729 / 7310 / +977 1 426 1979.
E-mail: moicppme@ntc.net.np

NETHERLANDS (CEPT, HAREC)
The Netherlands accepts both the CEPT and the HAREC, but if you have neither you can apply to the Agentschap Telecom for a visitor's licence. There is an application form available on the Agentschap Telecom website: click on "English", "Licences", then "Application temporary licence".

You will need a *certified* copy of your home licence.
Licensing authority: Agentschap Telecom;
website: www.agentschap-telecom.nl
Postal address: Postbus 450, 9700 AL Groningen, The Netherlands.
Street address: Emmasingel 1, 9726 AH Groningen.
Tel: +31 50 587 7444 (Monday to Friday 0800 - 1700 local time); **fax:** +31 50 587 7400.
E-mail: agentschaptelecom@at-ez.nl

NETHERLANDS ANTILLES (CEPT)
The Netherlands Antilles (the islands of Bonaire and Curacao off the northern coast of South America, and the northern Caribbean islands of Dutch Sint Maarten, St Eustatius and Saba) are one of the territories outside the CEPT area that nevertheless accepts the CEPT Licence.

If you do not have a CEPT Licence you may apply for a visitor's licence from the Bureau Telecommunicatie in Willemstad on Curacao. You may also apply for a special event callsign (e.g. for a contest) of the form PJnL, where n = 2 for Curacao, 4 for Bonaire, 5 = St Eustatius, 6 = Saba, 7 = Sint Maarten, and L is any single letter. The cost of the special event callsign is USD $15 for

Licensing office in Curacao, Netherlands Antilles.

PHOTO: PJ2T WEBSITE

the licence, plus $25 for the special event callsign, plus $9 in stamp charges and $11 bank fees, making a total of $60.

Contact the licensing officers (in 2008 they were Ms Omaira Verhey and Ms Ivy Ortega) to request an application form. You will need a *certified* copy of your home licence and the application should be received at least eight weeks before you arrive in the Netherlands Antilles.

Licensing authority: Bureau Telecommunicatie. *Postal address:* Director of Bureau Telecommunicatie, PO Box 2047, Willemstad, Curacao, Netherlands Antilles, Caribbean. *Street address:* Director of Bureau Telecommunicatie, Industrieterrein Groot Davelaar 139, Willemstad, Curacao; **Tel:** +599 9 463 1700, ext 713; **fax:** +599 9 736 5275.

NEW ZEALAND (CEPT, HAREC)

The Radio Spectrum Management (RSM) department of the Ministry of Economic Development handles amateur radio licensing in New Zealand.

Paragraph 4 of the so-called 'General User Radio Licence' which came into effect in 2006 states that: "Persons visiting New Zealand who hold a current amateur certificate of competency, authorisation or licence issued by another administration, may operate an amateur station in New Zealand for a period not exceeding 90 days, provided the certificate, authorisation or licence meets the requirements of Recommendation ITU-R M.1544 or CEPT T/R 61-01 or CEPT T/R 61-02 and is produced at the request of the chief executive... The callsign must be the national callsign allocated by the other administration to that person, in conjunction with the prefix or suffix 'ZL' which is to be separated from the national callsign by the character '/' (telegraphy), or the word 'stroke' (telephony)."

Thus visitors from almost any country (not just CEPT countries) can operate in New Zealand for a period of up to 90 days without the necessity of obtaining or applying for a licence.

The RSM website has a lot of useful information for radio amateurs in, or visiting, New Zealand. Click on to "Licensing", "Types of licences" then "Amateur licences". There is then a further link to the General User Radio Licence.

If you are staying in New Zealand for more than 90 days you may apply to RSM for a visitor's licence.

Licensing authority: Licensing Department, Radio Spectrum Management, PO Box 2847, Wellington 6140, New Zealand; **website:** www.rsm.govt.nz **Tel:** +64 3 962 2603 (Monday to Friday, except public holidays, 0830 - 1700 local time); **fax:** + 64 4 499 0797 **E-mail:** info@rsm.govt.nz

NICARAGUA

An application form (in Spanish) is available on the website of TELCOR, the licensing authority in Nicaragua. Go to "Formatos", then download the form at "Formato para el Servicio de Radioaficionados y Radioclubs" (the same form is available in two different places on the 'Formatos' web page).

Licensing authority: Instituto Nicaraguense de Telecomunicaciones y Correos (TELCOR), Edificio TELCOR, Av Bólivar, Esquina Diagonal a la Cancillería, Apartado 2664, Managua, Nicaragua; **website:** www.telcor.gob.ni **Tel:** +505 2 227 348 / 350; **fax:** +505 2 227 757. **E-mail:** info@telcor.gob.ni

NIGER

Amateur radio licences are issued to residents in Niger, and it is also possible, though more difficult, for a short-term visitor to obtain one.

Licensing in Niger is handled by the Autorité de Régulation Multisectorielle (ARM) and their website has an application form (in French) for Fixed, Mobile or Portable stations which, although not specifically for the Amateur Service, could probably also be used to apply for an amateur licence. Go to "Les Formulaires", and then click on "Autorisation Fréquence Réseau Radioélectrique Indépendant" under "Réseau Privé indépendant".

Licensing authority: Autorité de Régulation Multisectorielle (ARM), 64 Rue des Bâtisseurs, BP 13179, Niamey, Niger; **website:** http://niger.arm-niger.org **Tel:** +227 2073 9008 / 9011; **fax:** +227 2073 8591. **E-mail:** arm@arm-niger.org

NIGERIA

Visitor's licences are *only* granted to persons resi-

Nigeria, 5N, call districts.

dent in Nigeria for a minimum period of six months. Short-term operation is not permitted. All applicants must be current members of the Nigeria Amateur Radio Society (NARS) before a licence will be granted.

First contact NARS to apply for membership: Nigeria Amateur Radio Society [NARS], 5 6921A Road, Gwarinpa Housing Estate, PO Box 7502 Wuse, Zone 3, 900003, Abuja, FCT, Nigeria; tel: +234 9 672 5612 / +234 80 7724 2150; fax: +234 9 672 5612; e-mail: oyekunle.ajayi@ties.itu.ch (NARS Secretary-General: Oyekunle Bamidele Ajayi, ONA, 5N0OBA.

You will be sent details of how to join NARS and an application form for a licence. The whole procedure will take approximately three months. Note that you should not attempt to take any transmitting equipment into Nigeria until you have the licence and a customs clearance letter signed by the NARS Secretary-General.

Licensing authority: Federal Ministry of Information and Communications, Radio House, Herbert Macaulay Way (South), PMB 247, Garki, Abuja, Nigeria.

Tel: +234 9 234 4107; **fax:** +234 9 523 2834.

E-mail: festus.daudu@ties.itu.int (Eng Festus Yusufu Narai Daudu, Acting Director, Spectrum Management Department).

NIUE

You will have no problem to obtain a visitor's licence in Niue. Apply in person with the usual documents at the Post Office in the main square in Alofi.

Licensing authority: Director of Posts and Telecommunications, Post and Telecom Department, Government of Niue, Alofi, Niue.

Tel: +683 184.

NORWAY (CEPT, HAREC)

Amateur radio licensing in Norway is dealt with by the Norwegian Post and Telecommunications Authority (NPT).

Their website states, "A radio amateur who holds a valid CEPT licence according to CEPT recommendation T/R 61-01 may use his radio amateur equipment during short stays in Norway without having to apply for a separate licence. The callsign to be used is LA/home call. The same applies for Svalbard except that the prefix to be used is JW. Radio amateurs coming from countries which have not implemented the CEPT licence should contact the NPT at firmapost@npt.no The CEPT licence is not valid for Jan Mayen, Bouvet Island or Peter I Island."

Licensing authority: Post-og teletilsynet (PT); **website:** www.npt.no

Postal address: PO Box 93, 4791 Lillesand, Norway.

Street address: Nygård 1, Lillesand.

Tel: +47 22 824 600; **fax:** +47 22 824 640.

E-mail: firmapost@npt.no

OMAN

In order to receive an amateur radio licence in Oman it is necessary to be resident in the coun-

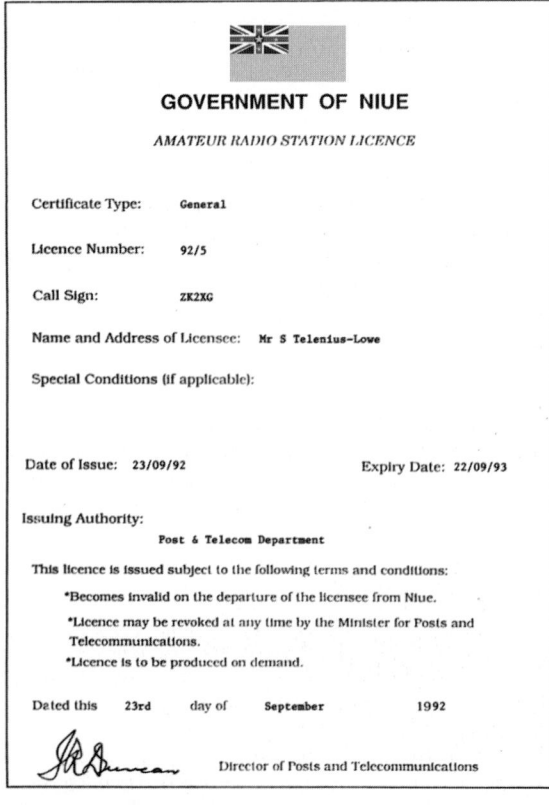

The Niue, ZK2, licence.

try; Oman will not issue a licence for you to use your own equipment on a short-term visit. It is, however, possible to obtain permission to operate the Royal Omani Amateur Radio Society (ROARS) club station, A47RS. ROARS also carries out the amateur radio examinations and is the *de facto* licensing authority in Oman on behalf of the Ministry of Communications.

For further information contact Royal Omani Amateur Radio Society, PO Box 981, Muscat PC 113, Sultanate of Oman; tel: +968 24 600 407 or +968 24 537 777; fax: +968 24 698 558; e-mail: roars@omantel.net.om

To request permission to use A47RS, write well in advance and enclose a copy of your home licence. ROARS HQ and the A47RS club station are located on top of a hill close to the Intercontinental Hotel.

Licensing authority: Oman Telecommunications Regulatory Authority (TRA), PO Box 579, Ruwi PC 112, Sultanate of Oman; **website:** www.tra.gov.om **Tel**: +968 24 574 303; **fax:** +968 24 565 464 **E-mail:** traoman@tra.gov.om

PAKISTAN

Amateur radio licences are only issued to Pakistani citizens, not to foreigners. Visitors are, however, allowed to operated from properly-licensed club stations, e.g. AP2ARS and AP2WAP. Contact the Pakistan Amateur Radio Society [PARS], PO Box 1450, Islamabad 44000, Pakistan; tel: +92 51 287 5099 / 287 6077 / 273 755; fax: +92 51 282 7581; e-mail: ap2nk@paknet2.ptc.pk *or* ap2nk26@hotmail.com (PARS President Nasir Khan, AP2NK).

Licensing authority: Pakistan Telecommunication Authority (PTA), PTA Headquarters F-5/1, 44000 Islamabad, Pakistan;
website: www.pta.gov.pk
Tel: +92 51 287 8145; **fax:** +92 51 287 8147.
E-mail: chairman@pta.gov.pk

PALAU

Amateur radio licensing in Palau is carried out by the Division of Transportation and Communication at the Bureau of Commercial Development, part of the Ministry of Commerce and Trade.

Capt Arvin Raymond, Chief of the Division of Transportation and Communication, Koror, Palau.

Their website states "Amateur radio operators who wish to operate in Palau will need to apply for an amateur radio licence from the Division of Transportation and Communication. Currently, ham licensing is a free government service. There are three classes of amateur radio privileges in Palau, namely novice, technician and general. To apply for an amateur radio licence, fill out the application form for amateur radio licence and mail or fax it to the Division of Transportation and Communication along with a copy of your amateur radio licence from your home country."

If you are staying at the Palau Pacific Resort (see Rental Stations section of this book), the Api Corporation in Japan can apply for your Palau licence on your behalf. Their fee is USD $70.00 and you should apply at least 40 days before your arrival in Palau. Send a copy of your amateur radio licence and / or certificate, a copy of your passport, give a choice of two-letter callsigns (T88**) and enclose a self-addressed envelope and International Reply Coupon. If you want a vanity callsign (T80 and single-letter suffix), the cost is USD $120.

Licensing authority: Ministry of Commerce and Trade, Division of Transportation and Communication, PO Box 1471, Koror, Palau 96940; **website:** www.palaugov.net/mincommerce/ tanscomm.html
Tel: +680 488 4343; **fax:** +680 488 3207.
E-mail: dot@palaunet.com

PALESTINE

It is possible for foreigners from some countries to receive a visitor's licence in the Palestinian Territories. The first criterion is that you must be from a country that has "friendly relations" with the Palestinian Authority. You must also have a licence from the country of which you are a citizen (licences are only granted on the basis of a foreign licence to citizens of that country: if, for example, you have an American licence but are not an American citizen you will not be able to receive a visitor's licence in Palestine).

An application form for an amateur licence has been uploaded to OH2MCN's 'Worldwide Information on Licensing for Radio Amateurs' website at www.qsl.net/oh2mcn/e4a.pdf

The completed application form should be sent to the Deputy Minister of Telecommunications in Gaza, Mr Zuhair Laham. Ensure the application is received well before your visit to the Palestinian Territories and collect the licence in person from the Ministry of Telecommunications office. Licences are valid for three months, at a cost of USD $30 or one year (USD $60) and the callsign granted for foreigners is E4/home call.

Licensing authority: Mr Zuhair Laham, the

Deputy Minister of Telecommunications, Ministry of Post & Telecommunications, Palestinian National Authority , Gaza, Palestine, via Israel.
E-mail: zlaham@marna

or

Mr Ahmad Qarout, Adviser, Ministry of Post & Telecommunications, Government Building, Al Ersal Street, Rammala, Palestine, via Israel.
E-mail: qarout@hotmail.com

PANAMA (IARP)

There are several ways of operating amateur radio when in Panama. Firstly, if you hold an International Amateur Radio Permit you may operate signing HPn/home call, where n is a digit representing the Panama call district from which you are operating. Before doing so, however, it is necessary to inform the Ministry of Government and Justice of the place and your dates of your operation.

Since 2006, Panama has extended the same privileges to CEPT Licence holders, again providing the Ministry of Government and Justice is informed of the place and dates before the operation starts.

If you are not from an IARP or CEPT Licence country you may apply for a Panamanian Temporary Operation Permit ('TOP'). To do so, you need to grant the power of attorney to a Panamanian lawyer to request the TOP on your behalf. You should provide the following: a photocopy of your home licence certified by a notary public *and* authenticated by a Panamanian Consulate; a photocopy of your passport certified by a notary public in the Republic of Panama; two identity photographs size 1 x 1 inch; USD $8 in fiscal stamps, obtained at any National Bank de Panama branch; and the licence fee of USD $20.

Finally, a full Panamanian licence is available for foreigners who are resident or retired in Panama. The procedure is the same as for the TOP.
Licensing authority: Department of Communications, Ministry of Government and Justice (Ministerio de Gobierno y Justicia / Medios de Comunicación Social);
website: www.gobiernoyjusticia.gob.pa
Postal address: Apartado Postal 1628, Panamá 1, Republic of Panama.
Street address: Former Howard Air Force Base, Building 703, 2nd floor.
Tel: +507 512 2165 / +507 212 2122 (Monday to Friday 0830 - 1630 local time); **fax:** +507 512 6001.
E-mail: informa@gobiernoyjusticia.gob.pa

PAPUA NEW GUINEA

Amateur radio licensing is efficient and quick in Papua New Guinea. Licensing is handled by the

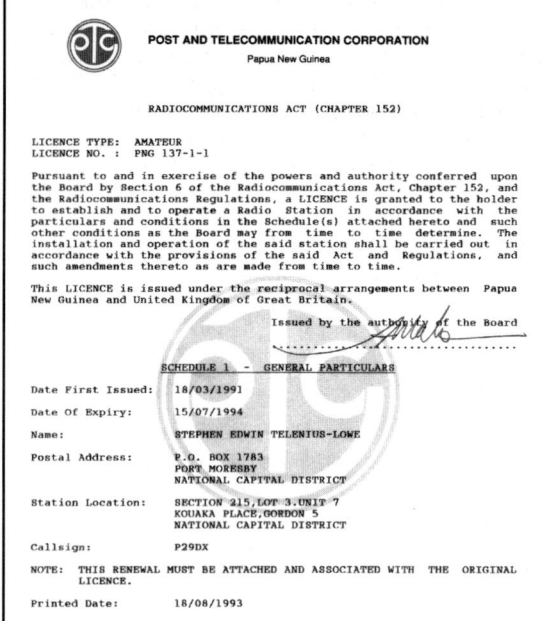

Papua New Guinea, P2, licence (issued by PTC, prior to PANGTEL).

Papua New Guinea Radiocommunication and Telecommunication Technical Authority (PANGTEL) and there is an application form for an amateur radio licence (form TR105) readily available on the PANGTEL website.

Fill in the form and mail to PANGTEL a few weeks before the licence is required. Upon arrival in the country, visit the office, pay the licence fee, and the licence will be available for collection. Both Boroko and Hohola are districts of Port Moresby, in the National Capital District (NCD).
Licensing authority: PANGTEL;
website: www.pangtel.gov.pg
Postal address: PANGTEL, PO Box 8444, Boroko, NCD, Papua New Guinea.
Street address: PANGTEL, Corner of Frangipani Street and Croton Street, Hohola, Port Moresby.
Tel: +675 325 8633;
fax: +675 325 6868 / +675 300 4829.
E-mail: uoome@pangtel.gov.pg

PARAGUAY

The IARU member society, RCP, can help visitors to obtain a permit to operate in Paraguay. Contact Radio Club Paraguayo (RCP), PO Box 512, Asuncion 1209, Paraguay; tel: +595 21 446124; fax: +595 21 446124; e-mail: zp5aa@ telesurf.com.py *or* zp5aa@zp5aa.org *or* hsbhsb@ gmail.com (IARU Liaison Officer Hernando Bertoni, ZP5HSB) for further details.

If you are staying with Tom, ZP5AZL (see 'Rental Stations' section), he will help you to obtain an operating permit: you should send him a copy of your licence and passport and USD $15.
Licensing authority: Ministerio de Obras Públicas y Comunicaciones, Oliva y Alberdí, Asunción 1221, Paraguay; **website:** www.mopc.gov.py
Tel: +595 21 414 9000 / 9711 / 9763;
fax: +595 21 414 9606.
E-mail: mopc@mopc.gov.py *or* comunicaciones@mopc.gov.py

PERU (CEPT, IARP)

Since Peru recognises both the CEPT Licence and the IARP, licensing should not present any difficulty for most amateurs from North and South America or Europe. When operating under the CEPT Licence or the IARP in Peru, you should add OA plus the relevant call district indicator *after* your callsign, e.g. M0QQQ/OA4. The map below shows the correct call district digits to use.

If you do not hold either a CEPT Licence or an IARP, contact the IARU member society for details of how to obtain a visitor's licence: Radio Club Peruano (RCP), Av Los Ruisenores Este 245, Urb El Palomar, San Isidro, Lima 27, Peru; tel:

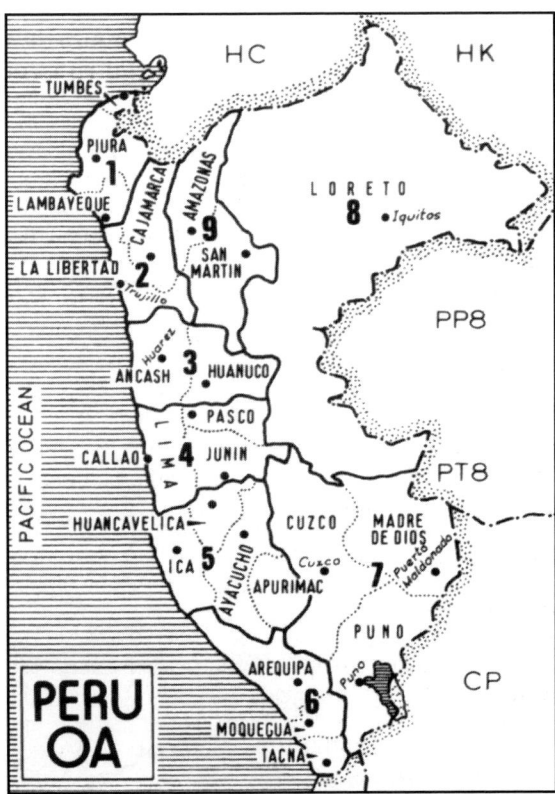

Peru, OA, call districts.

+51 1 224 0860; fax: +51 1 224 2792; e-mail: oa4amn@infonegocio.net.pe (IARU Liaison Officer Oscar Pancoruo, OA4AMN).
Licensing authority: Ministerio de Transportes y Comunicaciones, Jirón Zorritos No 1203, Lima 1, Peru; **website:** www.mtc.gob.pe
Tel: +51 1 615 7498 / 7800; **fax:** +51 1 615 7784.
E-mail: minist@mtc.gob.pe

PHILIPPINES

Amateurs from those countries that offer reciprocal privileges to Philippine amateurs (which does not include Japan) may apply for a visitor's licence, either direct to the National Telecommunications Commission (NTC) or through the Philippine Amateur Radio Association (PARA). Details of the NTC head office are given below; in addition there are at least 14 regional offices which are listed on the NTC website.

An 'Information Sheet' and a licence application form are available on the PARA website at www.para.org.ph by clicking on "Main Page" then "Licensing". As well as these two forms you will need copies of your passport and licence (translated to English if necessary), three colour photographs (1in x 1in), a certificate of good character issued by your embassy, the police or a church minister, the name and exact location of the place where you will operate, and make, model, serial numbers and frequency ranges of your equipment.

PARA will process your application with NTC. You are requested to join PARA: the licence application fee (approx PHP 1140 pesos, about GBP £13) includes membership fee. Further details from PARA, tel: +632 974 1986; fax: +632 249 1020; e-mail: dx1par@yahoo.com
Licensing authority (head office): Director, Radio Regulations & Licensing Department, National Telecommunications Commission (NTC), Bir Road, East Triangle, Diliman, Quezon City 1100, Philippines; **website:** www.ntc.gov.ph
Tel: +63 2 924 3787 / 4024 / 4042 / 4072 / 7128; **fax:** +63 2 924 4048 / 4072.
E-mail: rrld@ntc.gov.ph *or* ntc@ntc.gov.ph

PITCAIRN ISLANDS

A licence is issued without difficulty to *bona fide* visitors to Pitcairn Island. Apply well in advance of proposed visit.
Licensing authority: Pitcairn Islands Office, PO Box 105 696, Auckland, New Zealand.
Tel: +64 9 366 0186; **fax:** +64 9 366 0187.
E-mail: admin@pitcairn.gov.pn

POLAND (CEPT)

If you do not have a CEPT Licence, apply to the Polish Office of Electronic Communications (UKE) for a visitor's licence. Send a written application

Poland, SP, call districts.

with a copy of your home licence and a list of equipment and antennas that you plan to use. A one-year licence is issued with the SO prefix, call district digit and three letter suffix.

Licensing authority: Department of Frequency Resources Management, Urzad Komunikacji Elektronicznej (UKE, Office of Electronic Communications), 18/20 Kasprzaka Street, 01-211 Warsaw, Poland; **website:** www.uke.gov.pl
Tel: +48 22 534 9125 / 9156 (Monday to Friday 0815 - 1615); **fax:** +48 22 534 9155 / 9175.
E-mail: uke@uke.gov.pl

PORTUGAL (CEPT, HAREC)

Although Portugal accepts the CEPT Licence and the HAREC, confirmation of Morse code ability at a minimum speed of 50 characters per minute (10WPM) is required for HF bands operation.

You may also operate from the Azores and Madeira under the terms of the CEPT Licence. The prefix to be used is CT1 in mainland Portugal. CT3 in Madeira CT3 and in the Azores: CU1 (Santa Maria), CU2 (Sao Miguel), CU3 (Terceira), CU4 (Graciosa), CU5 (Sao Jorge), CU6 (Pico), CU7 (Faial), CU8 (Flores) and CU9 (Corvo).

If you do not hold a CEPT Licence, apply to ANACOM for a visitor's licence. This is valid for up to 30 days but may be renewed. The callsign is of the form CT1/home call. Apply with copies of your licence and passport.

If you are staying in Portugal for more than three months you may obtain a full Portuguese licence (callsign of the form CT1QQQ) either by submitting a HAREC or, if you have a residence permit and are from a country with which Portugal has a reciprocal agreement.
Licensing authority: Autoridade Nacional de Comunicações (ANACOM), Av José Malhoa 12, 1099-017 Lisbon, Portugal;
website: www.anacom.pt
Tel: +351 21 721 1000 (0900 - 1600);
fax: +351 21 721 1001.

QATAR

It is not normally possible for short-term visitors to obtain an amateur radio licence in Qatar, although it may be possible for foreigners who are resident in the country (particularly if you have 'connections'). Contact the Qatar Amateur Radio Society (QARS) for assistance: 82 Suhaim Bin Hamad Road, PO Box 22122, Doha, Qatar; tel: +974 447 7911; fax: +974 447 7955; e-mail: a71a@qatar.net.qa
Licensing authority: Supreme Council of Information and Communication Technology (ICT Qatar), Al Mirqab Tower, PO Box 23264, Doha, Qatar; **website:** www.ict.gov.qa
Tel: +974 493 5922; **fax:** +974 493 5913.

ROMANIA (CEPT, HAREC)

Amateur radio licensing is handled by the Romanian National Regulatory Authority for Communications and Information Technology (ANRCTI). Its website is in English as well as Romanian and has a link from the home page to "Amateur Radiocommunications" from where there are numerous links for downloading documents of in-

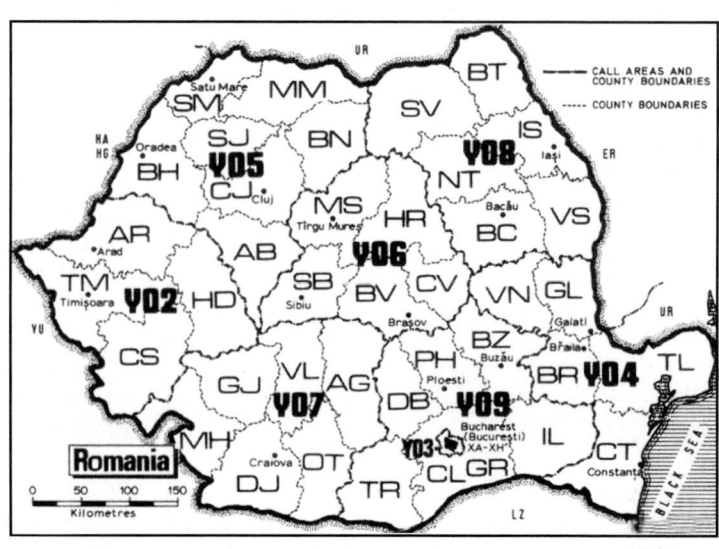

Romania, YO, call districts and county boundaries.

terest to radio amateurs.

Licensing authority: National Regulatory Authority for Communications and Information Technology (ANRCTI), 14 Libertatii Boulevard, Sector 5, 050706 Bucharest, Romania;
website: www.anrcti.ro
Tel: +40 21 307 5400; **fax:** +40 21 307 5402.
E-mail: anrcti@anrcti.ro

RUSSIA

It is possible to get a visitor's licence in Russia. Instructions and an application form are published on the ARRL website at www.arrl.org/FandES/field/regulations/io/russia-license-info-app.pdf

Applications should be made at least two months before the permit is required and it helps to have the assistance of a local radio amateur in the area from which you intend to operate, particularly for payment of the licence fee.

You could also ask the Russian Amateur Radio Union (IARU member society) for assistance: Soyuz Radiolyubitelei Rossii (SRR), PO Box 88, Moscow 119311, Russian Federation; tel: +7 495 485 4755; fax: +7 495 485 4981; e-mail: hqsrr@east.ru

Licensing authority: Glav Gos Svyaz Nadzor Rossii (GGSN), Dom 6, 2nd Spasonalivkovskij Pereulok, W-49, Moscow, GSP-1, 117951 Russia.
Tel: +7 495 973 1700 / +7 495 771 8400;
fax: +7 495 771 8734 / +7 495 292 5125 / +7 495 238 5102.

RWANDA

There were no amateur radio licences issued in Rwanda from 1994 until mid-2007. However, the present (2008) Director General of the Rwanda Utilities Regulatory Agency (RURA), Colonel Diogène Mudenge, has an interest in amateur radio and is licensed as 9X1AA. Several licences have been issued in 2007 and 2008 and a small national society, the Rwanda Amateur Radio Union (RARU), has been formed with the intention of becoming an IARU member society. 9X1 and 9X5 callsigns are issued to residents, and 9X0 callsigns to visitors.

It is suggested that applications for a licence in Rwanda be addressed to Col Mudenge.

Licensing authority: Rwanda Utilities Regulatory Agency (RURA), PO Box 7289, Kigali, Rwanda;
website: www.rura.gov.rw
Tel: +250 5 87066; **fax:** +250 5 87063
E-mail: mudenge@rura.gov.rw *or* dioge@rwanda1.com *or* dgoffice@rura.gov.rw

ST HELENA

Apply in person at the Post Office, situated in the main street of the capital, Jamestown, some 200 yards from the sea-front.

Licensing authority: The Postmistress, Post Office, Main Street, Jamestown, Island of Saint Helena, STH1ZZ, South Atlantic Ocean.
Tel: +290 2652;
fax: +290 2242.

St Helena's Post Office in Jamestown. ©Crown Copyright 2008.

ST KITTS & NEVIS

Amateur radio licences in St Kitts and Nevis are best obtained in person at the National Communications Regulatory Commission after arrival in the country. You will need to complete an application form, provide a copy of your home licence and passport and then you will be issued with an invoice for the licence fee of XCD EC$75 East Caribbean dollars (approx GBP £14.25). The invoice can be paid in cash with EC or US dollars at the Comptroller of Island Revenue office elsewhere in Basseterre. Then you will be issued with a receipt as proof of payment. Take the receipt back to the National Communications Regulatory Commission and the licence will be issued.

You should allow several hours for the whole process. Government offices are normally open 0800 - 1600 Monday to Friday.

All licences expire on 31 December each year (if you are visiting over the New Year holiday period you will therefore need to pay for two annual licences).

Licensing authority: National Communications Regulatory Commission.
Postal address: PO Box 1958, Basseterre, St Kitts and Nevis, West Indies.
Street address: Corner Wigley Avenue and Jones Street, Fort Land, Basseterre, St Kitts.
Tel: +1 869 466 6872.

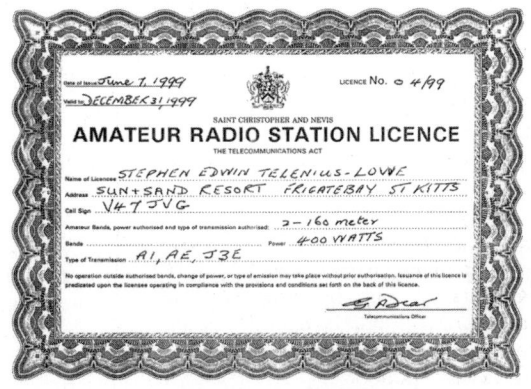
St Kitts and Nevis, V4, licence.

ST LUCIA

Amateur radio licensing is handled by the National Telecommunications Regulatory Commission (NTRC) of St Lucia. Their website has a link to an application form for an amateur radio or CB licence (under "Forms and Registers", "Applications", "Class Licence (Type B) Application Forms"). The form is also available as Annex III-C(iii) to a 71-page 'Application and Licensing Procedures' pdf document that is available from the "Applications" page of the website by clicking under "Related Links" on the right-hand side of the page.

You need to complete *three* copies of the application form and attach two passport-sized photographs, documentary proof that you have passed the Radio Amateur Examination (copy of licence or Amateur Radio Certificate) and photocopies of the technical specifications of the equipment. The completed application should be accompanied by the application fee of XFD EC$25 (East Caribbean dollars, approx GBP £4.75) or USD $10.

Licensing authority: National Telecommunications Regulatory Commission;
website: www.ntrc.org.lc
Postal address: Telecommunications Licence Application, The Secretary, National Telecommunications Regulatory Commission, PO Box GM 690, Castries, St Lucia.
Street address: National Telecommunications Regulatory Commission, Global Tile Building, Bois d'Orange, St Lucia.
Tel: +1 758 458 2035; **fax:** +1 758 4532558.
E-mail: ntrc_slu@candw.lc

ST VINCENT & THE GRENADINES

Amateur radio licensing is handled by the National Telecommunications Regulatory Commission (NTRC) of St Vincent & the Grenadines, and the licensing procedure and application form are virtually identical to those of St Lucia.

Go to the NTRC website at www.ntrc.vc and download the application form (the second form under "Applications" and "Class Licences").

As with St Lucia, you will need to complete *three* copies of the application form and attach two passport-sized photographs, documentary proof that you have passed the Radio Amateur Examination (copy of licence or Amateur Radio Certificate) and photocopies of the technical specifications of the equipment.

You should enclose an International Money Order for XFD EC$25 (East Caribbean dollars, approx GBP £4.75) or USD $10, payable to the NTRC. (This fee covers the application process; there are no licence fees.)

The NTRC will issue a letter which will allow easy processing of the radio equipment through customs. Depending on the timeframe involved, the NTRC will post the licence and letter to the visitor prior to arrival. If the timeframe is short they will fax copies to you and still mail the originals or you can collect the originals upon arrival.

Licensing authority: National Telecommunications Regulatory Commission (NTRC), KCCU Financial Centre, Granby Street, Kingstown, St Vincent and the Grenadines; **website:** www.ntrc.vc
Tel: +1 784 457 2279; **fax:** +1 784 457 2834.
E-mail: info@ntrc.vc

SAMOA

Licensing in Samoa is efficient and generally trouble-free. Apply by letter, fax or e-mail with *certified* copies of your home licence and passport, as well as specifications of your equipment, proposed location of operation in Samoa and dates of arrival and departure from Samoa.

If you intend to pick up the licence in person after arriving in Samoa, you should apply at least two weeks before the licence is required and you may pay the licence fee of WST $65 tala (approx GBP £13.15) when you pick up the licence. If the licence fee is included with the application, the licence can be posted to you, but in that case allow at least two months before the licence is required.

The licence is valid for a year and callsigns issued to visitors are of the format 5W0 plus two letters. You will be allocated your initials unless you specifically request another callsign.

Licensing authority: Attention Mr Fa'afetai Karanita Ah Kuoi, Monitoring and Spectrum Management Division, Ministry of Communications and Information Technology (MCIT);
website: www.mcit.gov.ws
Postal address: MCIT, Private Mail Bag, Apia, Samoa.
Street address: MCIT, 2nd Floor, CA & CT Plaza, Savalalo.
Tel: +685 26117 / 26709 (0800 - 1635, Monday to Friday); **fax:** +685 24671.
E-mail: fak@samoa.ws

SAN MARINO

San Marino has *not* implemented CEPT Recommendation T/R 61-01 and therefore the CEPT Licence is not valid in the Republic.

Also, it is not possible for foreigners to receive an HF visitor's licence in San Marino. Visitors may, however, be granted permission to use the ARRSM club station's antennas on the HF bands: see the 'Rental Stations' section of this book for further details.

If you have a CEPT Licence you may request a visitor's licence for the VHF bands from 144MHz and up (6m is not permitted). To do so, send a request and a copy of your licence to Mr Michele Giri, Telecommunications Collaborator, General Direction of Post & Telecommunications. The licence is valid for three months and there is a small licence fee involved.

Licensing authority: Direzione Generale, Poste e Telecomunicazioni, Contrada Omerelli 17, 47890 San Marino A-2, Republic of San Marino.
Tel: +378 0549 882555;
fax: +378 0549 882888 / 992760.
E-mail: dirposte@omniway.sm

SAO TOME & PRINCIPE

It is possible to get a visitor's licence in Sao Tome and Principe. The CEPT Licence is recognised, but some non-CEPT licences may also be accepted. You should apply well in advance of your visit with a letter written in Portuguese stating the dates you will be in the country, the location from which you wish to operate, and the bands and modes requested. You may request a specific callsign - give two or three suggestions in case your first choice has already been allocated - and enclose a copy of your licence.

The licence fee is 'negotiable' but should be around USD $25 - $30. You are advised to ensure that the licence is issued before your arrival: if you wait until you are in the country the licence fee may be several times this amount.

Licensing authority: Companhia Santomense de Telecomunicações (CST), Av 12 de Julho, Caixa Postal 141, São Tomé, Republica Democratica de São Tomé e Principé.
Tel: +239 2 22225 / 6;
fax: +239 2 21324 / 22500.

SAUDI ARABIA

It is not possible for short-term visitors to receive an amateur radio licence in Saudi Arabia. However, foreigners who have their "official residence" in the Kingdom may apply to the Communications and Information Technology Commission (CITC) for a licence, assuming they first hold a licence in the country of which they are a citizen.

Two separate licences are required: an operator licence and a station licence. The CITC website has application forms for both (in Arabic and English). Click on "English" then "Spectrum Management" and "Spectrum Management Services and Application Forms" to download the forms.

The full amateur radio regulations and technical conditions are also available for downloading from the same website page.

Note that you must not import any amateur radio equipment until you have a licence: customs will not release any radio equipment until they have obtained approval from CITC and after making sure that it conforms to the conditions and specifications approved by CITC.

Licensing authority: Communications and Information Technology Commission (CITC), PO Box 75606, Riyadh 11588, Kingdom of Saudi Arabia;
website: www.citc.gov.sa
Tel: +966 1 461 8003;
fax: +966 1 453 1289 / 461 8002.
E-mail: info@citc.gov.sa

SENEGAL

You should apply around four months before the licence is required with copies of your passport, home licence and a *handwritten* letter *in French* requesting the licence.

If you wish to import your own equipment into Senegal you will also need a transmitting certificate. To apply, once again a handwritten letter in French is required listing the make, model and serial number of all transmitting equipment, together with a photocopy of the technical specifications of the equipment.

The callsign issued is 6W followed by a call district digit followed by your home callsign, e.g. 6W1/M0QQQ. However, since 2006 a special single-letter callsign with the 6V prefix is also available, and these licences are issued strictly in alphabetical order, e.g. 6V1O, 6V1P etc.

The licence fee is EUR 55 euros (approx GBP £43) or 65 euros (approx £51) for a special callsign, plus 20 euros (approx £15.70) for each piece of transmitting equipment imported.

If you are staying at Le Calao (see 'Rental Stations' section of this book for details) they can take care of all formalities with the Senegalese authorities and obtain your callsign as well as your temporary transmitting certificate (if required).

Licensing authority: Agence de Régulation des Télécommunications et des Postes (ARTP Sénégal), Route des Almadies, Angle Djoulikaye, BP 14130, Dakar - Peytavin, Senegal;
websites: www.artp-senegal.org *and* www.artp.sn
Tel: +221 33 869 0369; **fax:** +221 33 869 0370.
E-mail: contact@artp.sn

SERBIA

Contact the Amateur Radio Union of Serbia for help in obtaining a visitor's licence: Amateur Radio Union of Serbia, Trg Republike 3/VI, PO Box 48, YU-11001 Belgrade, Serbia; tel / fax: +381 11 634 437; e-mail: yu0srs@beotel.yu
Licensing authority: Republic Telecommunication Agency (RATEL), Visnjiceva 8, 11000 Belgrade, Republic of Serbia;
website: www.ratel.org.yu
Tel: +381 11 322 0909 / +381 11 324 2673;
fax: +381 11 323 2537.
E-mail: ratel@ratel.org.yu

SEYCHELLES

There is an application form for a Seychelles amateur radio licence on OH2MCN's world-wide licensing website at www.qsl.net/oh2mcn/s79a.htm

The application form should be completed *in full* and in *triplicate* and submitted around two months before the licence is required with a copy of your home licence and a description of the equipment to be used.

There is an application fee of SCR R10 Seychelles rupees (approx GBP £0.65) plus a licence fee of R500 (approx GBP £32) for a three-month licence, or R1000 (approx GBP £64) for an annual licence. The licence fee must be paid before the licence can be issued. Without a licence you may not import equipment into the Seychelles.

You may request a special callsign at the time you submit your application.

Licensing authority: Mrs Mina Crea, Chief Executive Officer, Seychelles Licensing Authority (SLA), Orion Mall Building, PO Box 3, Victoria, Mahe, Seychelles; **website:** www.sla.gov.sc
Tel: +248 224 314 / +248 283 444 (enquiries can also be made by calling +248 283 444 ext 245, 246, 424 or 427); **fax:** +248 224 256.
E-mail: md@sla.gov.sc

SIERRA LEONE

To request a visitor's licence in Sierra Leone write a letter formally asking permission to operate. You should enclose a copy of your home licence and a letter attesting to your good character (e.g. from the police, your national amateur radio society or your local church).

IARU Member Society, the Sierra Leone Amateur Radio Society (SLARS), PO Box 10, Freetown, Sierra Leone; tel: +232 22 3335 (SLARS President Mrs Cassandra Davies, 9L1YL) may be able to help with your application.

The application should be sent to *both* the National Telecommunications Commission (NTC) *and* the Ministry of Information and Communications (see below).
Licensing authority: National Telecommunications Commission (NTC), 13 Regent Road, Hill Station, Freetown, Sierra Leone.
Tel: +232 22 236 858; **fax:** +232 22 235 791.
and Permanent Secretary, Ministry of Information and Communications, Ministerial Building, 5th Floor, George Street, Freetown, Sierra Leone.
Tel: +232 22 220 186; **fax:** +232 22 227 337.

SINGAPORE

It is not possible to get a licence in Singapore as a short-term visitor. The official explanation for this is the city state's small size and resulting high population density, which potentially makes TVI and other EMC issues more of a problem here than elsewhere. To obtain a licence you must reside in the country for a minimum of three months.

If you qualify and wish to apply for a licence you should obtain an application form from the InfoComm Development Authority of Singapore (IDA) and, in addition to the usual copies of your passport and current licence you will also need a copy of your Amateur Radio Certificate, a letter of approval from the owner of the premises for the installation of antennas, and diagrams of the proposed antenna installation.

All radio equipment must be approved by IDA before use.

The licence fee is SGD $50 Singapore dollars (approx GBP £18.60) for a one-year licence.
Licensing authority: InfoComm Development Authority of Singapore (IDA), 14-00 Suntec Tower Three, 8 Temasek Boulevard, Singapore 038988, Singapore; **website:** www.ida.gov.sg
Tel: +65 6211 1734 / 0888;
fax: +65 6211 2213 / 2220.
E-mail: info@ida.gov.sg

SLOVAK REPUBLIC (CEPT)

Slovakia recognises the CEPT Licence, but it is also possible to obtain a Slovak callsign (in the series OM9, followed by three letters) by applying in English or German to the Telecommunications Office of the Slovak Republic.

You will need the full address in Slovakia

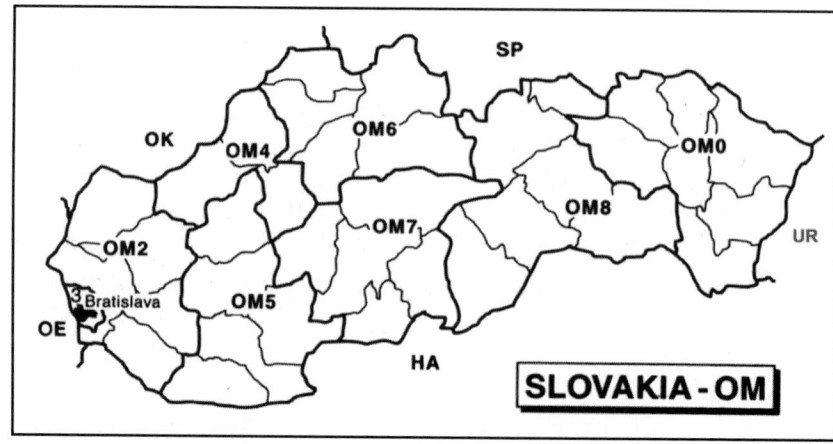

Slovakia, OM, call districts.

from where you are planning to transmit (this could be the address of a Slovak amateur you are visiting), a copy of your home licence, and a receipt for the licence fee. Check with either the Telecommunications Office or SARA, the IARU member society, how much the current licence fee is and how it should be paid. (Slovak Amateur Radio Association, SARA, Wolkrova 4, 85101 Bratislava; tel: +421 2 6224 7501; fax: +421 2 6224 7501.)

Licensing authority: Telekomunikacny urad Slovenskej republiky (Telecommunications Office of the Slovak Republic), Tovarenska 7, PO Box 18, 81006 Bratisalava 16, Slovak Republic; **website:** www.teleoff.gov.sk
Tel: +421 2 5788 1111; **fax:** +421 2 5293 2096.
E-mail: frequency@teleoff.gov.sk

SLOVENIA (CEPT, HAREC)

Slovenia accepts both the CEPT Licence and the HAREC, but if you have neither you should apply to the licensing authority, APEK, for a visitor's licence.

Members of the IARU national society ZRS may be able to help with such applications: contact Zveza Radioamaterjev Slovenije (ZRS), PP 180, SI-1001 Ljubljana, Slovenia; tel: +386 1 252 2459; fax: +386 1 422 0422; e-mail: zrs-hq@hamradio.si
Licensing authority: Head of Frequency Management and Licensing Department, Post and Electronic Communications Agency of the Republic of Slovenia (APEK), Stegne 7, PO Box 418, SI-1001 Ljubljana, Slovenia; **website:** www.apek.si
Tel: +386 1 583 6360 / 6300 / 6313 / 6314 (office hours Monday and Wednesday 0900 - 1100);
fax: +386 1 511 1101.
E-mail: info.box@apek.si

SOLOMON ISLANDS

There are no difficulties involved in getting an amateur radio licence in the Solomon Islands. The easiest way is to go in person to the Ministry of Infrastructure Development Spectrum Management Division office in Honiara. Spectrum Management is located in a blue building on top of Vavaya Ridge next to the National Parliament building.

Take your passport and home licence, and photocopies of both, with the licence fee of SBD $50 Solomon Islands dollars (approx GBP £3.60) and your H44 licence should be issued while you wait.

Note about operating from Temotu Province (Santa Cruz Islands): Licences for Temotu Province (the Santa Cruz Islands) may also be obtained from the Spectrum Management office in Honiara. Even if you already have an H44 licence for the rest of the Solomon Islands, you will still need to obtain a separate H40 licence for Temotu, and will

need to pay the licence fee for both.
Licensing authority: Spectrum Management Division, Ministry of Infrastructure Development, PO Box G8, Honiara, Solomon Islands.
Tel: +677 25888; **fax:** +677 28054.
E-mail: spectrum@solomon.com.sb

SOMALIA

Before discussing amateur radio licensing in Somalia, it is necessary to describe the geographical and political make-up of the country. The country internationally recognised as the Republic of *Somalia*, with Mogadishu as its capital, was formed in 1960 from the former British Somaliland Protectorate and Italian Somaliland.

Following a period of civil war in Somalia, in May 1991 the part of Somalia that was formerly British Somaliland declared itself independent as the Republic of *Somaliland*, with Hargeisa as its capital, although its independence has never been recognised by any country or international organisation. Both Somalia and Somaliland are made up of a number of states and there are conflicting border disputes both between the two republics and between some of the states that make up the two territories.

One such state is *Puntland*, in the north-eastern part of Somalia. In 1998 its leaders declared it to be the autonomous state of Puntland, although they did not seek complete independence from Somalia. Despite the ongoing border disputes, 'Puntland State of Somalia' is now a relatively stable part of the country and suffers somewhat less from the civil unrest that is still taking place in

Map showing Puntland and the other de facto states of Somalia.

southern Somalia.

While it would be foolhardy to attempt to operate amateur radio in southern Somalia, and anyway probably virtually impossible to obtain a licence from the authorities that are internationally recognised in Mogadishu (detailed below, for what it's worth), it *is* possible to receive a permit to operate in Puntland. Sam Voron, VK2BVS, has written a series of amateur radio regulations for the state, and he conducts amateur radio courses and examinations for residents of Puntland. Licences with 6O0 callsigns are available for short-term visitors to the state and these are accepted, by the ARRL DXCC desk at least, as legitimate Somali licences.

To obtain permission to operate in Puntland send an e-mail to Sam Voron, VK2BVS, at somaliahamradio@yahoo.com with copies to kajooje@yahoo.com (Mohamed Yasin Isak, 6O0MY) *and* hasan2jeer@hotmail.com (Hassan Mohamed Jamma, 6O0XJ). Attach scans of your home licence, passport photo identification page *and* front cover, and details of the date and time of your arrival at Gaalkacyo airport, and the flight number. You will be met at the airport with both your entry permission and your visitor's licence. You should bring with you *two* copies of your licence and passport pages and two identification photographs. There is an amateur radio club station at the Radio Gaalkacyo broadcast station. (Note: Gaalkacyo is also spelt Galkacyo, Galcaio, Galkao or Galkayo.) For further details contact Sam Voron, VK2BVS / 6O0A, tel: +61 2 9417 1066; e-mail: somaliahamradio@yahoo.com

Licensing authority (official): Ministry of Posts and Telecommunications, Directorate General, Mogadishu, Somalia.
Tel: +252 1 270 379 / +252 1 222 588;
fax: +252 1 251 800.
E-mail: mpt@globalsom.com

SOUTH AFRICA (CEPT)

Short-term visitors with a CEPT Licence may operate without further ado. If you are to be resident in South Africa you may apply to the Independent Communications Authority of South Africa (ICASA) for a full South African callsign. Short-term visitors without a CEPT licence should also apply to ICASA for a visitor's licence (ZS, followed by the call district digit followed by /own call).

Send a copy of your passport and home licence, with proof of current validity, along with an application listing your full name, address and telephone number in South Africa, and date of arrival in and departure from the country.

Licensing authority: Independent Communications Authority of South Africa (ICASA);
website: www.icasa.org.za

Postal address: Private Bag X10002, Sandton 2146, Republic of South Africa.
Street address: Blocks A, B, C and D, Pinmill Farm, 164 Katherine Street, Sandton, Johannesburg.
Tel: +27 11 566 3000 / 3001;
fax: +27 11 444 1919 / +27 11 321 8577.
E-mail: info@icasa.org.za

SOUTH GEORGIA & SOUTH SANDWICH ISLANDS

South Georgia and the South Sandwich Islands (SGSSI) form a separate UK Overseas Territory with a Commissioner as head of the government (prior to 1985 they had been part of the Falkland Islands Dependencies). Those wishing to operate amateur radio from South Georgia or the South Sandwich Islands should note that, in addition to the VP8 licence (which can be obtained in the Falkland Islands) it is also necessary to obtain a permit from the Commissioner, SGSSI, for all visits to the islands.

Applications should be made by completing the downloadable form on the South Georgia Government website at www.sgisland.org (individual tourists or visitors on cruise ships, yachts or expeditions need not complete the form providing their organisers have done so). Travel to the islands should not be undertaken without having first obtained official approval. Completed visitor application forms should be returned to the Office of the Commissioner no later than 60 days before the intended visit.

Most visits to the islands are ship-based and overnight stays on land are not normally permitted. Any visit to the islands which involves an overnight stay is defined as an "expedition" and specific permission must be obtained.

The South Georgia Government team is based at Government House in Stanley in the Falkland Islands. The Operations Manager (in 2008, Mr Richard McKee) addresses tourism and expeditions as well as managing fishery patrols. Contact him at Office of the Commissioner, Government House, Stanley, Falkland Islands, South Atlantic; tel: +500 28200; fax: +500 28201; e-mail: gov.house@horizon.co.fk

Licensing authority: Superintendent of Posts and Telecomms, c/o Post Office, Stanley, Falkland Islands, South Atlantic.
Tel: +500 27180.
E-mail: adminpost@townhall.gov.fk

SPAIN (CEPT)

If you do not hold a CEPT Licence you should apply to the Dirección General de Telecomunicaciones for a visitor's licence. The website at www.mityc.es/Telecomunicaciones has information (in Spanish) of interest to radio amateurs: go to "Secciones", "Espectro Radioeléctrico", then

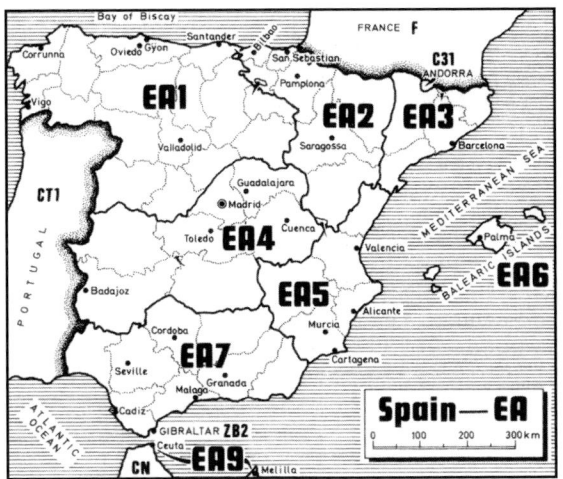

Spain, EA, call districts.

"Radioaficionados y CB-27".

Note that the Spanish licence allows for operation from all Spanish territory: the Balearic Islands, EA6; the Canary Islands, EA8; and Spanish North Africa (Ceuta and Melilla), EA9, as well as mainland Spain (EA1 - 5 and EA7).
Licensing authority: Dirección General de Telecomunicaciones y para la Sociedad de la Información, Área de Concesiones y Autorizaciones (Radioaficionados), Capitán Haya 41, 28071 Madrid, Spain;
website: www.mityc.es/Telecomunicaciones
Tel: +34 91 346 1500;
fax: +34 91 346 1520 / 2229.

SRI LANKA
At times, it has been difficult to obtain a visitor's amateur radio licence in Sri Lanka but at the time of compiling this book (mid-2008) they are being issued again. Visitors' licences are issued in the series 4S7xxG: three-letter callsigns with the last letter being 'G' for 'Guest', and the licence is valid for up to one year.

There are guidelines and an application form for a visitor's licence on the Telecom Regulatory Commission (TRC) webpage at www.trc.gov.lk - go to "Spectrum Management" then "Amateur Radio License". In addition to the completed application form you will need a covering letter requesting permission to operate in Sri Lanka, giving the local address from where you plan to operate and the dates of your stay; copies of your home licence and passport; a police report or other document certifying your good character; the names, addresses and signatures of two Sri Lankan guarantors or referees (may be, but not necessarily, radio amateurs), photocopies of the technical specifications of your transmitting equipment; and two identification photographs 1.75in x 1.5in in size.

The application should be sent to the Director General of Telecommunications, who will check it and then forward it to the Ministry of Defence, whose permission is required to issue a licence. This process takes time and so the earlier the application can be filed the better, with a minimum time of two months before the licence is required. The applicant is informed when Defence clearance has been received and the licence will then be issued on your arrival in the country upon payment of the licence fee of LKR 500 Sri Lanka rupees (approx GBP £2.35).
Licensing authority: Director General of Telecommunications, Telecommunications Regulatory Commission of Sri Lanka (TRC), 276 Elvitigala Mawatha, Colombo 8, Sri Lanka;
website: www.trc.gov.lk
Tel: +94 11 268 3843 / 9345;
fax: +94 11 268 9341.
E-mail: dgtsl@trc.gov.lk

SUDAN
It can be difficult and time-consuming to obtain an amateur radio licence in Sudan. You are advised to request help from members of the Sudanese Amateur Radio Association (SARA) if you seek to obtain a licence in Sudan. Contact Dr Sid Ahmeed, ST2SA, PO Box 1533, Khartoum, Sudan; e-mail: sidst2@gmail.com or Dr Nader Ali Omer, ST2NH, e-mail: st2nh@yahoo.com or e-mail SARA at sudanham@yahoo.com

A staff member at the National Telecommunication Corporation (NTC), Eng Hassab Elrasoul Aboulgasim, ST2AA, who is also an Under Secretary at the Ministry of Information and Communications, has attended the IARU 'Amateur Radio Administrative Course for Regulators' at the African Advanced Level Telecommunications Institute in Nairobi, Kenya. He is publicly thanked on the SARA website "for his recognition of the importance of ham radio as a means of progress and creation of engineers for the country. He has started issuing licences and regulated the ham radio for future developments."

Radio Scouting is popular in Sudan and is supported by both the NTC and SARA. You may improve your chances of gaining a licence if you are able to help with the Radio Scouting movement in Sudan in some way.
Licensing authority: National Telecommunication Corporation (NTC), Technical Administration, Frequency Planning; **website:** www.ntc.org.sd
Postal address: PO Box 2869, Khartoum 11111, Sudan.
Street address: 21 Street intersection with Mohamed Najib Alamarat Street, Khartoum.

Tel: +249 183 484 490 / +249 183 562 374 /
+249 183 483 203;
fax: +249 183 483 202 / +249 183 562 352 /
+249 183 484 489.
E-mail: itisalat@ntc.org.sd *or*
itisalat@sudanmail.net

SURINAME

Licensing is handled by the Telecommunicatiebedrijf Suriname (TELESUR), a department
of the Ministry of Transport, Communications and
Tourism of Suriname.

To apply for a licence send a written request
with a copy of your home licence, the precise address in Suriname from where you intend to operate, and the dates you will be in the country.
You may request a specific callsign. Below are
given the addresses of the TELESUR Radio Monitoring office, TELESUR's head office in
Paramaribo, and the Ministry of Transport, Communications and Tourism.

If you are staying at the location of Ramon
Kaersenhout, PZ5RA (see 'Rental Stations' section
of this book), he may be able to help you to get the
visitor's licence. Alternatively, contact the IARU
member society for help: Vereniging van Radio
Amateurs in Suriname (VRAS), PO Box 566,
Paramaribo, Suriname; tel: +597 471951 / 400880;
fax: +597 474 114; e-mail: vdhcrown@sr.net
Licensing authority: Telecommunicatiebedrijf
Suriname (TELESUR), Directeur Telesur Bureau,
Radio Controle en Monitoring, Cultuurtuin,
Paramaribo, Suriname.
Tel: +597 545 022
or TELESUR Centrum, Heiligenweg 14,
Paramaribo, Suriname; **website:** www.telesur.sr
Tel: +597 473 944 / 474 242; **fax:** +597 421 919.
E-mail: struiken@sr.net
or Ministry of Transport, Communications and
Tourism, Prins Hendrikstraat 26 - 28, Paramaribo,
Suriname; **website:** www.mintct.sr
Tel: +597 420 422 /423; **fax:** +597 420 425.
E-mail: secmin@mintct.sr

SWAZILAND

It is not difficult to obtain a visitor's licence in
Swaziland. The Chairman of the Radio Society of
Swaziland (RSS), Willie Long, 3DA0BD, will apply for a licence on behalf of visiting amateurs.
Send a copy of your licence and passport personal
details pages to Willie. Note that as there is only
dial-up Internet / e-mail access in Swaziland at
present, the licence and passport should be sent
by fax, not e-mail.

You may request a specific callsign suffix.
Licensing authority: c/o Chairman, Radio Soci-
ety of Swaziland (RSS), Willie Long, 3DA0BD, PO
Box 349, Malkerns, Swaziland.

Tel: +268 602 8080; **fax:** +268 416 2048.
E-mail: long@posix.co.sz

SWEDEN (CEPT)

The Swedish Post and Telecom Agency PTS (Post-
och telestyrelsen) handed over responsibility for
amateur radio licensing to the national society SSA
(Sveriges Sändareamatörer) on 1 October 2004. In
the first instance therefore you should contact
SSA for any queries about licensing in Sweden.
Details of both organisations are given below.
Licensing authority: Foreningen Sveriges

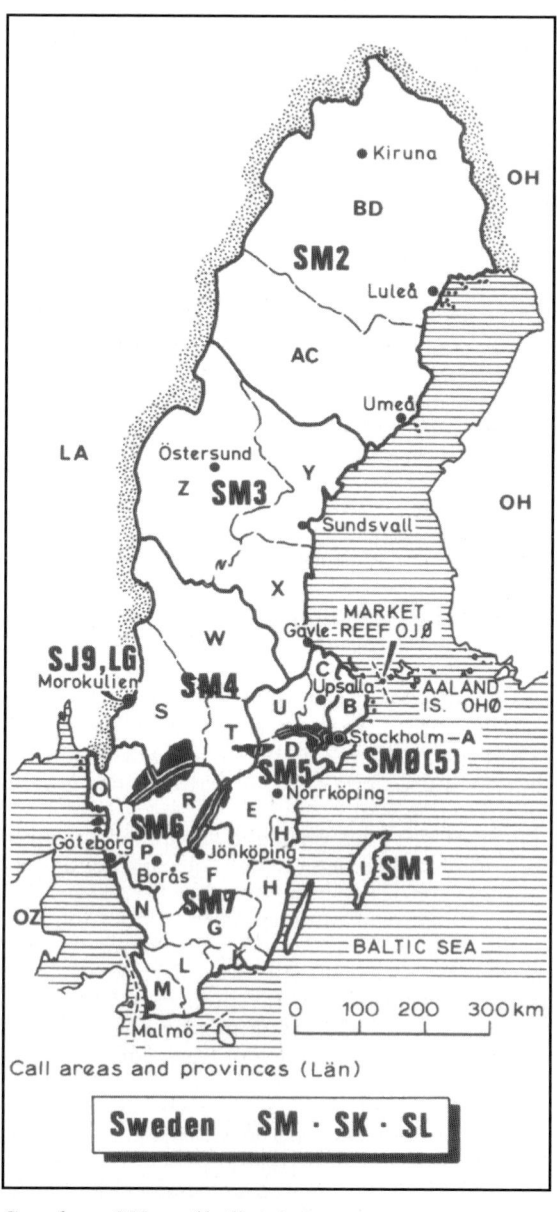

Sweden, SM, call districts.

Sändareamatörer (SSA); **website:** www.ssa.se
Postal address: PO Box 45, SE-191 21 Sollentuna, Sweden.
Street address: Turebergs Alle 2, Sollentuna.
Tel: +46 8 5857 0273 / +46 8 5001 1173; **fax:** +46 8 5857 0274.
E-mail: hq@ssa.se
or
PTS (Post- och telestyrelsen);
website: www.pts.se
Postal address: Box 5398, SE-102 49 Stockholm, Sweden.
Street address: Birger Jarlsgatan 16, Östermalm, Stockholm.
Tel: +46 8 678 5500 (Monday to Friday 0800 - 1700); **fax:** +46 8 678 5505.
E-mail: pts@pts.se

SWITZERLAND (CEPT, HAREC)
Switzerland accepts both the CEPT Licence and the HAREC. If you have neither, contact OFCOM in order to obtain a visitor's licence.
Licensing authority: Federal Office of Communications (OFCOM), 44 Rue de l'Avenir / Zukunftstrasse 44, PO Box, CH-2501 Biel-Bienne, Switzerland; **website:** www.bakom.ch
Tel: +41 32 327 5511 (Monday to Friday 0730 - 1200 & 1330 - 1700); **fax:** +41 32 327 5555.
E-mail: info@bakom.admin.ch

SYRIA
It is not easy for an individual to obtain an amateur radio licence in Syria. However, a few visitor's licences have been issued to groups during the last few years.

The following appears on the website of the IARU member society, the Syrian Scientific Technical Amateur Radio Society, SSTARS: "Individual foreigners can get a licence to operate amateur radio in Syria if there is a mutual agreement between the Syrian government and the other government recognising each other's amateur radio licences. When such an agreement exists, the foreign radio amateur should apply to SSTARS, sending a copy of his amateur licence and a photocopy of his passport. SSTARS will then follow up on the matter with the Syrian Telecommunications Establishment (STE) for issuing a provisional licence. The licensed foreign amateur will then be allowed to practice at the club station of SSTARS. DXpeditions from well-known radio societies / clubs can be licensed as a group." (See www.qsl.net/tir/Home.htm)

The best contact is the president of SSTARS, Dr Omar Shabsigh, YK1AO, PO Box 245, Damascus, Syria; tel: +963 11 311 4540 / 612 1279 / 231 8796; fax: +963 11 311 4540; e-mail: shabs.om@scs-net.org

Licensing authority: Syrian Telecommunications Establishment (STE), Directorate General, HQ Building, Fayez Mansour St Mazzeh Autostrade, PO Box 35108, Damascus, Syria.
Tel: +963 11 224 0300; **fax:** +963 11 612 0000 / +963 11 612 1208 / +963 11 224 2000.
E-mail: ste-gm@net.sy

TAIWAN (REPUBLIC OF CHINA)
Visitor's licences are issued through the IARU member society, the Chinese Taipei Amateur Radio League (CTARL). Contact CTARL and ask for an application form to be sent to you.
Licensing authority: Chinese Taipei Amateur Radio League (CTARL), Room B2-22, No 6, Sec 1, Sinsheng S Road, Da-an District, Taipei City 106, Taiwan; **website:** www.ctarl.org.tw
Tel: +886 2 2322 1841; **fax:** +886 2 2322 5925.
E-mail: hq@ctarl.org.tw

TAJIKISTAN
Tajikistan is probably the easiest of the former Soviet Union countries in Asia in which to obtain a visitor's amateur radio licence. Applications are made via the country's national society TARL (Tajik Amateur Radio League).

Apply with a letter indicating the dates of your visit, your address in Tajikistan and enclose a photocopy of your home licence. Send the documents at least one month before your arrival in the country to: TARL Secretary, Nodir Tursoon-Zadeh, EY8MM, PO Box 303, Dushanbe 734001, Tajikistan; tel / fax: +992 372 212 844 / 847; e-mail: tarl@qsl.net When you arrive in Tajikistan you should make arrangements to meet an officer of TARL to pay the licence fee of USD $15 and pick up your licence. For short-term visits the callsign given is EY, followed by the call district, followed by your home call, e.g. EY8/M0QQQ for operation from Dushanbe, the capital. For visits of a year or more, a full EY callsign is given.
Licensing authority: Ministry of Transport and Communications, Aini 14, Dushanbe 734042, Tajikistan.
Tel: +992 372 211 713;
fax: +992 372 211 766 / +992 372 212 003.
E-mail: mt-rt@mail.ru

TANZANIA
It can take time, but if you have an overseas amateur radio licence you should be able to obtain a visitor's licence in Tanzania. The procedure is as follows. First, you need to *purchase* an application form from the Tanzania Communications Regulatory Authority (TCRA) for TZS 10,000 Tanzania shillings (approx GBP £4.25). Complete the form and submit it with a letter of application containing details of the dates and location of your

proposed operation in Tanzania and the make, model and serial number of your equipment. Enclose copies of your home licence, passport, and the technical specifications of your transceiver. Finally, you should enclose a letter or letters of recommendation, e.g. from your national amateur radio society and / or a police certificate of good conduct. When the paperwork has been processed you will need to make a further payment of 10,000 shillings in order to receive the callsign and licence document.

All this takes time and is much more easily achieved if you are in Dar-es-Salaam and can visit the TCRA office in person. Alternatively, a local contact can obtain the form and help the procedure along on your behalf.

If you do not know anyone in Tanzania, the Tanzania Amateur Radio Club (TARC) may be able to help. Contact TARC at PO Box 2958, Dar-es-Salaam, Tanzania; tel: +255 22 215 0942 / +255 22 215 0174 ext 266; fax: +255 22 215 2504. The club may also be able to provide a letter of recommendation as required by TCRA in the application process. The club requests a small donation if they are able to help you to obtain a licence.

Licensing authority: The Director General, Tanzania Communications Regulatory Authority (TCRA); **website:** www.tcra.go.tz
Postal address: PO Box 474, Dar-es-Salaam, Tanzania.
Street address: Mawasiliano House, Plot 304, Ali Hassan Mwinyi / Nkomo Road, Dar-es-Salaam.
Tel: +255 22 211 8947 / 8952 / +255 744 720 411; **fax:** +255 22 211 6664.
E-mail: dg@tcra.go.tz

THAILAND

Licensing in Thailand is somewhat complicated, to say the least! The easiest option is to obtain a permit to operate HS0AC, the club station of the Radio Amateur Society of Thailand (RAST) near Bangkok. These permits are available to amateurs from any country, but they are valid for operating the HS0AC station only. Contact the station manager, Finn Jensen, OZ1HET, by e-mail at oz1het@rast.or.th for further details.

If you wish to operate any other station (including your own equipment) from Thailand, you must be a citizen of one of only seven countries with which Thailand has finalised reciprocal licensing arrangements: Austria, Germany, Luxembourg, Sweden, Switzerland, UK and USA. Belgium, Canada, Norway and Spain are believed to be in the process of setting up reciprocal agreements, but at the time of compiling this book (mid-2008) they had not been concluded and it is still not possible for citizens of those countries to obtain their own licence in Thailand.

If you are from one of the seven countries you may obtain your own licence, but there are several stages involved. The first (and also easiest and quickest) is to obtain an *Operator Licence*. This is the licence with which a callsign is associated. Reciprocal licences are in the series HS0Z followed by two more letters and the callsigns are issued in alphabetical order. This operating licence, however, only allows you to operate an existing station in Thailand with your own callsign: it does *not* allow you to set up your own station, nor to import equipment.

If you wish to take your own equipment into Thailand, you will need an *import permit*. Import permits are required for *all* transceivers, including VHF / UHF handhelds (it is illegal to possess such equipment without having the proper documentation).

Once the equipment has been legally imported and cleared by customs you may apply for an *Equipment Licence*. The equipment must be inspected by the Posts and Telegraph Department (PTD) of the National Telecommunications Commission (NTC) and registered with them before the equipment licence can be issued.

Even then the equipment can only be used at a location which has a *Station Licence*. This requires a further application form from the PTD, a copy of your *Operator Licence* and *Equipment Licence*, written permission from the owner of the property that the installation and operation of the amateur radio station and its antennas are permitted, a copy of the property owner's identity card, a copy of the house registration certificate and a detailed map showing the location of the house / apartment.

Given all the above it is perhaps not surprising that in practice generally only long-term residents of Thailand are able to set up their own stations! However, many foreign amateurs hold-

The Thai Operator Licence takes the form of a credit-card sized ID card.

The NTC building in Bangkok.

ing the HS0Z *Operator Licence* are able to get on the air from club stations or as guest operators from residents' stations.

RAST has extremely comprehensive information on all the various stages required to obtain a full reciprocal licence on its English-language website at www.qsl.net/rast/text/LicensingSept05.html You may also download reciprocal licence application forms from www.qsl.net/rast/text/Recip.htm

RAST is entrusted by the NTC to screen applicants and issue a letter of support for foreign licence applications as well as to register formally the application in accordance with the Thai government's regulations. As such, it is necessary to join RAST if you intend to take out a reciprocal licence: contact RAST's International Affairs officer, Tony Waltham, HS0ZDX, c/o Radio Amateur Society of Thailand (RAST), GPO Box 2008, Bangkok 10501, Thailand; tel: +66 2 618 4435 / +66 2 392 8672; fax: +66 2 618 4435 / +66 2 240 3665; e-mail: tonyw@inet.co.th

Licensing authority: National Telecommunications Commission (NTC) Secretariat, Amateur Radio Licensing Department, Posts and Telegraph Department, 87 Soi Sailom Soi 8, Phaholyothin Road, Phayathai, Bangkok 10400, Thailand; **websites:** http://eng.ntc.or.th (in English), www.ntc.or.th (Thai).
Tel: +66 2 271 3511 / +66 2 272 7054 / +66 2 278 0151; **fax:** +66 2 278 1736 / +66 2 290 5240.
E-mail: inter_org@ntc.or.th

TIMOR-LESTE (EAST TIMOR)
A small number of 4W6 licences has been issued to visitors since the United Nations Transitional Administration in East Timor (UNTAET) pulled out and handed over full independence to the country in 2002.

Licensing is handled by ARCOM (Autoridade Reguladora das Comunições), a division of the Ministry of Transport and Communications. An ARCOM radio licence application form (intended primarily for VHF PMR mobile, repeater and base stations, but which can also be used for amateur radio applications) can be found on OH2MCN's website at www.qsl.net/oh2mcn/4wa.pdf

One-month or one-year licences are available and have been issued by Sr Jose Maria de Araujo (Spectrum Manager) and approved by Sr Nicolau Santos Celestino, the Head of ARCOM. ARCOM's English-language website has a section on Radio Communications and Licensing but this was 'under construction' at the time this book was being compiled.

Licensing authority: Autoridade Reguladora das Comunições (ARCOM), Av Bispo de Modeiros, Dili, Timor-Leste; **website:** www.arcom.tl
Tel: +670 333 1163 / 9330 / 9341 / 9343;
fax: +670 333 9339.
E-mail: ARCOM-info@arcom.tl

TOKELAU
Although Tokelau is an overseas territory of New Zealand, all travel to the islands is by the *MV Tokelau*, or her sister ships the *Samoa Express* and *Lady Naomi,* from Apia in Samoa. It is necessary to book accommodation and to obtain a visitor permit from the Tokelau Apia Liaison Office (TALO) in Apia before departing Samoa. TALO can also assist you to obtain a ZK3 visitor's licence in Tokelau.

Licensing authority: Tokelau Apia Liaison Office (TALO).
Postal address: Tokelau Apia Liaison Office, PO Box 865, Apia, Samoa.
Street address: Tokelau Apia Liaison Office, Fugalei Street, Apia.
Tel: +685 20822 / 20823, +685 71805, +685 777 1096 / 1807; **fax:** +685 21761.
E-mail: zakp@lesamoa.net, maka@lesamoa.net, mitingauchun@lesamoa.net

TOGO
It is possible to obtain a visitor's licence in Togo. Applications should be made in French, and preferably in person at the office of the Autorité de Régulation des Télécommunications (ART&P) in Lomé. The application and licence fee is EUR 55.00 euros (approx GBP £44).

Licensing authority: Autorité de Réglementation des Secteurs des Postes et Télécommunications (ART&P); **website:** www.artp.tg
Postal address: Boite Postale 358, Lomé, Togo.
Street address: 9th Floor, Immeuble BTCI Building, 169 Bvd du 13 Janvier, Lomé.
Tel: +228 222 8385 / 8422 / +228 226 6809;
fax: +228 222 8612.
E-mail: artp@artp.tg

TONGA

It is easy to get a visitor's licence in Tonga, but you must present your *original* home licence and passport along with photocopies to the Licensing Officer in the Prime Minister's Office in Nuku'alofa. You will also need to complete an application form and pay the licence fee of TOP 20.00 Tonga Pa'anga (approx GBP £5.50).

You may request a specific callsign but be aware that many A35 two-letter callsigns have already been issued, so give a choice of three or four possible calls. You will be issued with a callsign and a receipt and may start to operate the same day. The licence document itself will be posted to you but this may take several weeks.

Licences are valid for one year and all have a common renewal date of 1 July. The maximum power permitted in Tonga is 100 watts.

Licensing authority: Radio Licensing Officer, Department of Communications, Prime Minister's Office; **website:** www.pmo.gov.to
Postal address: PO Box 62, Nuku'alofa, Tonga.
Street address: Ground Floor, Old Prime Minister's Office Building, Taufa'ahau Road, Nuku'alofa (near General Post Office).
Tel: +676 24644, **fax:** +676 23888.
E-mail: lpalefau@pmo.gov.to (Mr Lopeti Palefau).

TRINIDAD & TOBAGO (IARP)

The International Amateur Radio Permit is accepted in Trinidad and Tobago and holders of this permit may operate as 9Y/own call without any further paperwork being involved.

If you are not from an IARP country, you should apply to the Telecommunications Authority of Trinidad and Tobago (TATT) for a visitor's licence. An application form is available for downloading from the TATT website by clicking on "Forms" and then scrolling down the page to Form L2. The application form requires details of the manufacturer, model, serial number and date purchased of all equipment.

Note too that TATT needs both *originals* and copies of your home licence and passport. It is therefore suggested that you apply in advance with photocopies, and take your original home licence and passport to the TATT office after arrival in the country, from where it should be possible to pick up your Trinidad and Tobago licence. However, if you are flying directly to Tobago and not intending to visit Trinidad, this may be inconvenient, so you should check to see whether copies certified by a Notary Public are acceptable *in lieu* of original documents.
Licensing authority: Telecommunications Authority of Trinidad and Tobago (TATT), Suites 3-5, BEN Court, 76 Boundary Road, San Juan, Trinidad, Republic of Trinidad and Tobago;

website: www.tatt.org.tt
Tel: +1 868 675 8288; **fax:** +1 868 674 1055.
E-mail: info@tatt.org.tt

TRISTAN DA CUNHA

There is no problem for *bona fide* visitors to this, the most remote inhabited island in the world, to obtain a ZD9 licence. Enquiries should be made to the Administrator's Office.
Licensing authority: Superintendent of Posts and Telegraphs, The Administrator's Office, Edinburgh of the Seven Seas, Tristan da Cunha TDCU 1ZZ, Atlantic Ocean, via Cape Town South Africa.
E-mail: enquiriestdc@gmail.com

TUNISIA

It is not possible to for visitors to receive their own amateur radio callsign licence or callsign in Tunisia. However, visitors may obtain permission to operate the existing club stations in the country - see the 'Rental Stations' section of this book for further details or contact the International Liaison Officer of the Association Tunisienne des Radioamateurs (ASTRA), Mustapha Landoulsi, DL1BDF, Westlinteler Weg 30, 26506 Norden, Germany; tel / fax: +49 4931 12519; e-mail: info@landoulsi-norden.de
Licensing authority: Ministère des Technologies de la Communication, 3 bis Rue d'Angleterre, 1000 Tunis, Tunisia.
Tel: +216 71 323434 / 359000;
fax: +216 71 322686 /352353.
E-mail: communications@ministeres.tn

TURKEY (CEPT, HAREC)

CEPT Licence holders are allowed to operate in Turkey in accordance with CEPT Recommendation T/R 61-01 for a maximum period of three months.

The Turkish IARU member society, TRAC, has produced three documents written in Turkish which you are advised to print out and show to customs officers, along with your own licence, if requested to do so when temporarily importing your equipment into Turkey. The documents are available on the TRAC website at www.trac.org.tr - click on "Foreign Radio Amateurs (English)" to download them.

Amateurs who do not have a CEPT Licence or who plan to stay for longer than three months should apply for a one-year visitor's licence. Contact the Turkish Telecommunications Authority, Telekomünikasyon Kurumu (TK) or TRAC for an application form.

You will also need copies of your home licence and passport plus two passport-size photographs.

The visitor's licence callsign is TA followed

Turkey, TA, call districts.

by the call district digit followed by 'Z', followed by another letter, e.g. TA3ZX. TRAC may be contacted at PO Box 699, TR-80005 Karakoy, Istanbul, Turkey; tel: +90 532 376 5707 / +90 532 556 3396; fax: +90 212 257 7856 / +90 216 423 2350 / +90 216 386 5129; e-mail: hq.trac@trac.org.tr

Licensing authority: Telekomünikasyon Kurumu, Yesilirmak Sokak No 16, Demirtepe 06430, Ankara, Turkey; **website:** www.tk.gov.tr
Tel: +90 312 294 7125 / 7200 / +90 312 550 5314; **fax:** +90 312 294 714 / 7155.
E-mail: ird@tk.gov.tr

TURKS & CAICOS ISLANDS
New legislation for amateur radio was finalised at a meeting between members of the Turks and Caicos Amateur Radio Society (TACARS) and officials from the Ministry of Communications and the Attorney General's Office in January 2004. The legislation was enshrined in law in the Wireless Telegraphy (Amateur Radio Operator Licensing) Regulations 2004, which designated TACARS as the certifying organisation for amateur radio in the Turks and Caicos Islands. In other words, TACARS now has the responsibility of carrying out amateur radio licensing on behalf of the Ministry of Communications.

Visitors receive a licence with the callsign VP5/home call, but both resident and visitor licensees may request a VP5 or VQ5 single-letter Special Event callsign for a specific operating event. Requests for specific callsigns will be honoured if possible. Every licence issued is valid for the calendar year in which it is granted and expires on 31 December in that year (except for Special Event licences).

An amateur radio licence application form in pdf format can be found on the TACARS website at www.tacars.org/license.pdf Complete the form and send it with a copy of your current home licence to the Secretary of TACARS, Jody Millspaugh, VP5JM. The licence fee is USD $35.00, or $70.00 for a Special Event licence. You may send a US check (cheque), but if you do not hold a US bank account please confirm with Jody how the licence fee should be paid. You are advised to send the application by FedEx to: Jody Millspaugh, Cherokee Road, Providenciales, Turks and Caicos Islands, BWI. If you have sufficient time before the licence is required you may send the application by air mail post to: Jody Millspaugh, PO Box 218, Providenciales, Turks and Caicos Islands, BWI, but you should allow eight to 10 weeks for mail delivery. Please confirm by e-mail to jody@tciway.tc that the application, licence copy and check have been sent by either FedEx or mail.

The maximum power level permitted in the Turks and Caicos Islands is 1.5kW PEP.

When entering the islands it is recommended that you have two copies of a complete list of the equipment you are taking, indicating make, model number and serial number for each item, including antennas and other ancillary equipment. Customs officials may ask for this list. It is important to state, if asked, that the equipment will be returning with you when you leave. You may have to pay significant import duties for any items that do not leave the country on your departure. If you are bringing items that *will* stay in the country, you should be prepared to declare the value of the items (have invoice copies) and pay import duties. On departing the islands you may be asked to produce the list of equipment as evidence that the items are returning.

Licensing authority: Turks and Caicos Amateur Radio Society (TACARS), c/o Jody Millspaugh, PO Box 218, Providenciales, Turks and Caicos Islands, BWI; **website:** www.tacars.org
Tel: +1 649 946 4436 between 6.00pm and 7.00pm EST (2200 - 2300GMT).
E-mail: jody@tciway.tc

TURKMENISTAN
Amateur radio operation has been suspended by the government of the Republic of Turkmenistan since mid-2006, therefore it is currently impossible to obtain a licence. Even local amateurs who were licensed before the ban came into effect are at present off the air.

Licensing authority: Ministry of Communications, Zhitnikova Street 36, Ashgabat, Turkmenistan.
Tel: +993 12 352 153; **fax:** +993 12 390 420.
E-mail: mincom@telecom.tm

TUVALU

There is no problem to receive a visitor's licence in Tuvalu. Apply on arrival in Funafuti with the usual documents.
Licensing authority: Ministry of Communications, Transport and Tourism, Private Mail Bag, Vaiaku, Funafuti, Tuvalu.
Tel: +688 20 052 / 055 / 721;
fax: +688 20 722 / 800.
E-mail: tap@tuvalu.tv

UGANDA

It is not difficult to get a visitor's licence in Uganda, though it may take some time as the licence must be approved at a Board meeting of the Uganda Communications Commission (UCC), which normally sits only once a month.

Request an application form from the UCC Licensing and Standards Department and send it with a covering letter and a copy of your home licence. The licence fee is USD $60.
Licensing authority: Mr Simon Bugaba, Officer in Charge, Licensing and Standards, Uganda Communications Commission (UCC);
website: www.ucc.co.ug
Postal address: UCC, PO Box 7376, Kampala, Uganda.
Street address: UCC, 12th Floor, Communications House, Plot 1 Colville Street, Kampala.
Tel: +256 41 434 8830 / 8831 / 8835;
fax: +256 41 434 8832.
E-mail: ucc@ucc.co.ug *or* striples@ucc.co.ug

UKRAINE (CEPT)

Although Ukraine accepts the CEPT Licence, proof of Morse code proficiency is required for use of the HF bands.

If you do not hold a CEPT Licence apply to the Ministry of Transport and Communications at least two months before the licence is required with a covering letter, a copy of your home licence and a letter of recommendation from your national amateur radio society.

The Ukrainian Amateur Radio League (UARL) may be able to help with the application. Contact them at UARL PO Box 56, Kiev-1, 01001 Ukraine; tel: +380 44 457 0972 / +380 44 280 4649; fax: +380 44 457 7195; e-mail: ut2ub@ham.kiev.ua (President Andrej Lyakin, UT2UB).
Licensing authority: Ministry of Transport and Communications, 22 Khreschatyk Street, 01001 Kiev, Ukraine; **website:** www.stc.gov.ua

Tel: +380 44 278 1500;
fax: +380 44 278 6141 / +380 44 226 2673.
E-mail: mailbox@stc.gov.ua

UNITED ARAB EMIRATES

For many years it was impossible for anyone other than nationals of the UAE to receive an amateur radio licence here. In recent years, however, a small number of licences have been issued to foreigners who are resident in the Emirates. It is believed to be still impossible to receive a licence unless you have a residence or work permit.

Apply to the Telecommunications Regulatory Authority (TRA) in person. The head office is in Abu Dhabi but details are also given for their regional office in Dubai.
Licensing authority: Telecommunications Regulatory Authority (TRA), PO Box 26662, Abu Dhabi, United Arab Emirates; **website:** www.tra.ae
Tel: +971 2 621 2222 / +971 2 626 9999 (Saturday to Thursday 0800 - 1300);
fax: +971 2 621 2227 / +971 2 611 8229.
E-mail: spectrum@tra.ae
or
Dubai Office, Telecommunications Regulatory Authority, PO Box 116688, Dubai, United Arab Emirates.
Tel: +971 4 428 8888; **fax:** +971 4 428 8800.

UNITED KINGDOM (CEPT, HAREC)

It is easy to get a visitor's amateur radio licence in the UK. The CEPT Licence is accepted for visits of up to three months. For longer stays, amateurs from CEPT administrations which have adopted T/R 61-02 can apply to the Office of Communications (Ofcom) for a Full UK licence. You will need to complete a normal Amateur Radio Licence application form and to have a HAREC from a country that has implemented CEPT T/R 61-02.

Amateurs from countries which have not implemented CEPT Recommendation T/R 61-01 or T/R 61-02 will need to obtain a temporary UK licence from Ofcom. Such licences are issued on the strength of their home licence if a reciprocal agreement has been entered into between the UK and the government of their country. If the visitor is unable to provide a UK address (that is, he will be operating mobile) or where the length of stay is under six months, his temporary UK licence will be valid for up to six months. This licence is not renewable.

If the foreign radio amateur is a regular visitor with a permanent contact address in the UK, a Full UK licence (and callsign) may be issued on the basis of a reciprocal arrangement.

The correct prefixes to use before your home callsign when operating under the terms of the CEPT Licence or with a six-month visitor's li-

The British Isles, showing the correct prefixes to be used with the CEPT and visitors' licences.

cence are: M (England), MD (Isle of Man), MI (Northern Ireland), MJ (Jersey), MM (Scotland), MU (Guernsey and Dependencies), and MW (Wales) and *not* G, GD, GI, GJ, GM, GU and GW.

For those granted Full callsigns, the prefix should be changed when they operate from a different part of the country. For example, a visitor with the Full callsign of MM0QQQ at his address in Scotland should change his callsign to MW0QQQ if he operates from Wales.

Further information about amateur radio licensing in the UK can be found on the RSGB's website at www.rsgb.org/operating and from Ofcom.

Licensing authority: Office of Communications (Ofcom); **website:** www.ofcom.org.uk/licensing *Postal address:* Amateur and Maritime Team, Ofcom Licensing Centre, PO Box 56373, London SE1 9SZ, United Kingdom.
Street address: Riverside House, 2A Southwark Bridge Road, London SE1 9HA.
Tel: +44 20 7981 3131; **fax:** +44 20 7981 3333.
E-mail: licensingcentre@ofcom.org.uk

Uruguay, CX, call series letters.

URUGUAY (IARP)
Uruguay accepts the IARP. Those from non-IARP issuing countries should apply for a visitor's licence from URSEC. If you are renting any of Ron Serván's, CX2AQ, stations (see 'Rental Stations' section of this book), Ron can help you to obtain a CX licence.
Licensing authority: Unidad Reguladora de Servicios de Comunicaciones (URSEC), Avenida Uruguay 988, Montevideo 11200, Uruguay; **website:** www.ursec.gub.uy
Tel: +598 2 902 8082;
fax: +598 2 900 5708 / +598 2 901 2252.
E-mail: info@ursec.gub.uy

USA (CEPT, IARP)
Licensing in the USA is handled by the Federal Communications Commission (FCC). It should be noted that the FCC no longer issues reciprocal permits for non-US citizens (so-called 'aliens'). Although the USA accepts both the CEPT Licence and CITEL's IARP, since February 1999 the licensing regime has been relaxed further. Now non-US citizens from *all* countries that have a reciprocal licensing agreement with the USA (over 80 countries) may operate from the USA and its territories without having to apply for a licence or permit. Note that this applies to *non*-citizens of the USA only: an American citizen with an overseas licence may *not* operate in the USA under this scheme. Similarly this permission becomes invalid for an alien living in the USA should he take out US citizenship. A full list of countries with which the US has a reciprocal licensing

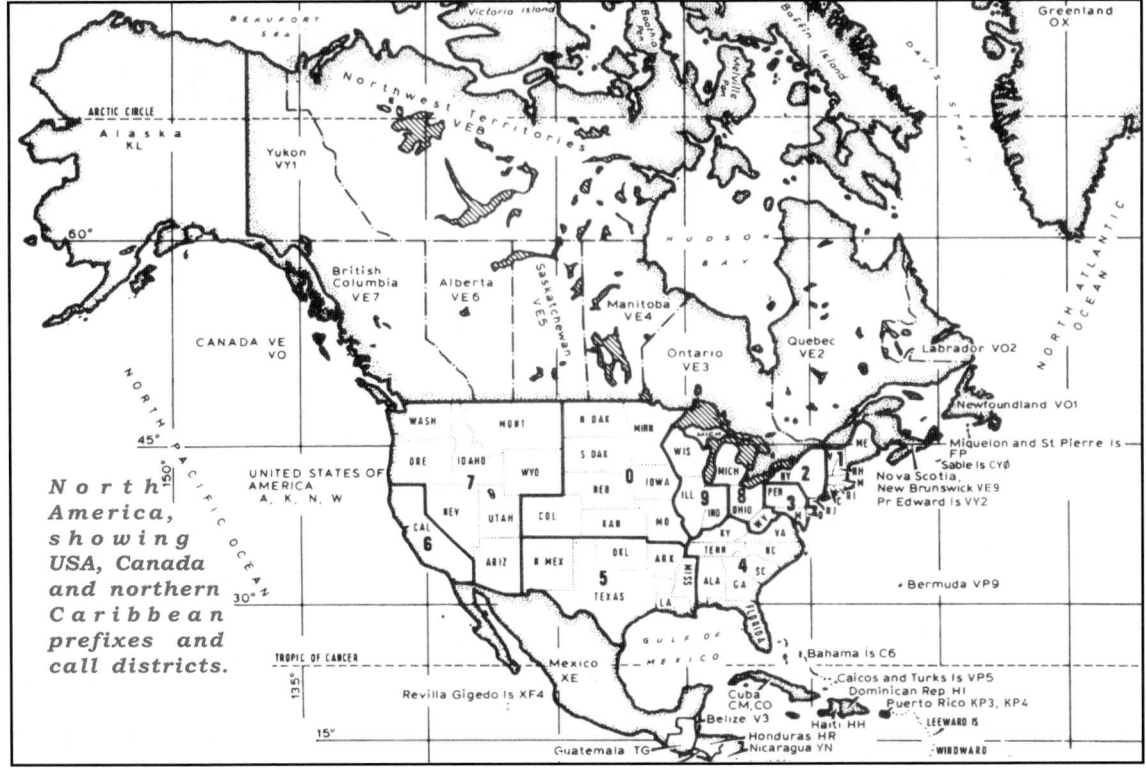

North America, showing USA, Canada and northern Caribbean prefixes and call districts.

agreement can be found on the ARRL website at www.arrl.org/FandES/field/regulations/io/recip.html

The callsign you should use is the "representative prefix" of the call area where you are located, followed by your home call, e.g. W1/M0QQQ, KH6/M0QQQ etc. There is an exception for Canadian amateurs, who should sign their own call first, followed by the US call district, e.g. VE3QQQ/W2.

Anyone who can provide the Federal Communications Commission (FCC) with a US mailing address can take the US amateur examinations and so obtain a full American licence, whether or not they are from a country which has a reciprocal licensing agreement with the USA. For further information contact the Volunteer Examiner Department at ARRL headquarters. The US mailing address could be that of an American amateur who is willing to forward mail on to you, if you do not have your own address in the USA.

There is a mass of information about licensing in the US on both the ARRL website at www.arrl.org (click on "Licensing") and on the FCC's Wireless Telecommunications Bureau (WTB) website at http://wireless.fcc.gov (scroll down the WTB's home page and click on "Amateur" under "Wireless Services").

Licensing authority: Federal Communications Commission (FCC), 445 12th Street SW, Washington DC 20554, United States of America; **websites:** www.fcc.gov *and* http://wireless.fcc.gov
Tel: +1 202 418 2150;
fax: +1 202 418 0398 / 1208.
E-mail: fccinfo@fcc.gov

UZBEKISTAN

No up to date information on amateur radio licensing is presently available. For those travelling to Uzbekistan who wish to take out a licence, the following is the contact information of the licensing authority, the Communications and Information Agency of Uzbekistan, as given on their own website.

Licensing authority: Department of Licensing and Standardisation, Communications and Information Agency of Uzbekistan, 28A Alisher Navoiy Street, Tashkent 100011, Uzbekistan;
website: www.aci.uz
Tel: +998 71 238 41 07 /08 / 37 / 39 / 78 / 79 or 85; **fax:** +998 71 239 8782 / +998 71 233 1695.
E-mail: info@aci.uz, deplicense@aci.uz *or* gkrch@aci.uz

VANUATU

It is not difficult to get a visitor's licence in Vanuatu, but it is best to go along in person to the

Telecom office in Port Vila after your arrival in the country. Take along *originals* and copies of your home licence and passport. The visitor's licence is of the type YJ0A followed by two further letters, e.g. YJ0AXX. The licence fee is about VUV 600 vatu (approx GBP £3.25).

Licensing authority: Managing Director, Telecom Vanuatu Ltd, PO Box 146, Port Vila, Vanuatu;
website: www.tvl.net.vu
Tel: +678 22185; **fax:** +678 22628.
E-mail: telecom@tvl.net.vu

VATICAN CITY STATE

It is not possible for individuals to obtain an amateur radio licence for use within the Vatican.

However, there are a couple of amateur radio club stations established in the Vatican City and extraterritorial enclaves belonging to the Vatican and it *may* be possible to obtain permission to operate from one of these. It is suggested you contact the respective station managers directly in order to ask permission (HV0A, c/o Francesco Valsecchi, IK0FVC, Via Bitossi 21, 00136 Roma (RM), Italy, *or* HV50VR, c/o Alex Carletti, IW0GPN, CP 123, 33085 Maniago (PN), Italy; e-mail: iw0gpn@tiscali.it)

Licensing authority: Governatorato, Secretariat General, Vatican City 00120, Vatican.
Tel: +39 6 6988 3158 / 3418;
fax: +39 6 6988 2954 / 5299.

VENEZUELA (IARP)

Venezuela accepts the IARP. Those from non-IARP issuing countries should apply for a visitor's licence from CONATEL.

The IARU member society, RCV, may be able to help visitors to obtain a permit. Contact them at: Radio Club Venezolano (RCV), PO Box 2285, Caracas 1010-A, Venezuela; tel: +58 212 781 4878 / +58 212 793 5404; fax: +58 212 793 6883; e-mail: rcv@radioclubvenezolano.org

Licensing authority: Comisión Nacional de Telecomunicaciones (CONATEL), Las Mercedes, Av Veracruz con Calle Cali, 1060 Caracas, Venezuela;
website: www.conatel.gob.ve
Tel: +58 212 993 6122 / +58 212 909 0336;
fax: +58 212 993 5389.
E-mail: conatel@conatel.gob.ve

VIETNAM

It is difficult, though not impossible, to receive an amateur radio licence in Vietnam. The main difficulty, for most visitors, is due to the 'Catch-22' that you cannot apply for the licence until you have a Vietnam entry stamp in your passport, but then the licence is only valid for the duration of your Vietnamese visa. Since most tourist visas are only of two weeks duration and it can take several days to a week for the licence to be issued, few short-term visitors are able to obtain one.

A further complication is that the application form is entirely in Vietnamese, so it is almost essential to have local help when applying for a licence.

Those wanting to try should first obtain the application form from the Ministry of Information and Communications and then submit the completed form with a covering letter (English is acceptable) requesting a licence, and copies of your passport personal details pages, Vietnam visa and Vietnam entry stamp in your passport, as well as your amateur radio certificate.

Note that what is required is an amateur radio *certificate* and *not* a copy of your home licence: generally an amateur radio licence document is *not* accepted.

The Vietnam Amateur Radio Club (VARC) may

Venezuela, YV, provinces and call districts.

HOME OFFICE

AMATEUR RADIO CERTIFICATE

This is to certify that

has taken the Radio Amateur Examination and the Post Office Morse Test and has qualified in the following subjects for the award of the Amateur Radio Certificate:

Elementary theory of radio communication
Knowledge of transmitting techniques
Knowledge of operating procedure appropriate to radio amateurs
Sending and receiving morse at the rate of 12 words a minute

Date 16 MAY 19   Signed

NOTE: This Certificate will not be regarded as exempting the holder from having to pass the Morse Test again if he applies for an Amateur (Sound) Licence more than twelve months after the date on which he passed the Post Office Morse Test mentioned above.

No. 744

MP&TI SEC/1238   (2881) 4306842/150683 8/74 1m ACS 821

The UK Amateur Radio Certificate. This is what is required if you wish to apply for a Vietnamese amateur radio licence: a certificate (not an overseas licence).

be able to help with the licence application procedure. Contact them at 49 Ben Chuong Duong Street, District I, Ho Chi Minh City, Vietnam; tel: +84 8 821 2193 / +84 8 829 4912; fax: +84 8 914 1008; e-mail: dtu@saigon.net.vn *or* ngbacaixv2a@yahoo.com (President: Nguyen Bac Ai, XV2A / 3W6AR).
Licensing authority: Ministry of Information and Communications (MIC), 18 Nguyen Du Street, Hanoi 10000, Vietnam; **website:** www.mic.gov.vn
Tel: +84 4 822 6622 / 9267 / +84 4 556 3464 / +84 4 943 0204; **fax:** +84 4 943 5296 / +84 4 822 6590 / +84 4 556 3458.
E-mail: dic@mic.gov.vn, banbientap@mic.gov.vn *or* tt_tt@mic.gov.vn

YEMEN

It is generally impossible for visitors to obtain an amateur radio licence in Yemen. There are also currently no Yemeni nationals with amateur radio licences.

The last licence that was issued to a foreigner was to Pekka Ahlqvist, OH2YY, who was active from Yemen as 7O/OH2YY for eight days in May 2002.

It is just possible that if you are resident in Yemen for a lengthy period of time, e.g. on a work contract, you *may* be able to obtain a licence. If you wish to try, the following are believed to be the current contact details of the radio regulatory authority.
Licensing authority: Ministry of Telecommunications and Information Technology, Airport Road, Al-Jiraf Sana'a, PO Box 25237, Sana'a, Yemen.
Tel: +967 1 331 452; **fax:** +967 1 331 457 / 469.
E-mail: dmcom@yemen.net.ye

ZAMBIA

It is not difficult for foreign residents to obtain an amateur radio licence in Zambia. It may be more difficult for short-term visitors, but members of the Radio Society of Zambia may be able to help. Contact them at Radio Society of Zambia (RSZ), PO Box 20332, Kitwe, Zambia; tel: +260 2 224 690 / +260 1 289 087; fax: +260 1 288 009; e-mail: ramsch@zamnet.zm
Licensing authority: Office of the Controller, Communications Authority of Zambia (CAZ); **website:** www.caz.zm
Postal address: CAZ, PO Box 36871, Lusaka 10101, Zambia.
Street address: CAZ, Plot 3141, Corner of Buyantanshi and Lumumba Roads, Lusaka.
Tel: +260 211 246 696 / +260 211 241 236 / +260 211 246 702; **fax:** +260 211 246 701.
E-mail: info@caz.zm, caz@zamnet.zm, jtembo@caz.gov.zm *or* nnankonde@caz.gov.zm

ZIMBABWE

The current political situation in Zimbabwe is not conducive to amateur radio licensing, although there is a small number of resident amateurs who have held licences for many years. It is unlikely that any new licences are being issued at present (mid-2008) but it is to be hoped that the situation will change for the better in the future. Anyone intending to apply for a licence in Zimbabwe is strongly advised to contact the Zimbabwe Amateur Radio Society (ZARS) in advance: ZARS, c/o Eric Christer, Z21FO, 8 Silwood Close, Chisipite-Harare, Zimbabwe; tel: +263 4 490 995; e-mail: christer@mweb.co.zw
Licensing authority: Manager, Special Telecommunications Services, Post and Telecommunications Corporation (PTC), Box CY331, Causeway, Harare, Zimbabwe.
Tel: +263 4 791 701; **fax:** +263 4 731 094.

Rental Stations Directory

In this section all the stations around the world that are available for rent by radio amateurs are listed: at least, all those that we know about! As noted in Chapter 1, the owner or operator of the stations listed in this section of the book were contacted, firstly in order to obtain their permission for the stations to be included in the book and, secondly, to ensure that all the details were correct at the time of publication.

The process of contacting the station owners revealed a number of anomalies. Firstly, there were several stations which are offered for rent on the Internet where the owners did not respond to my e-mails enquiring about their properties. The e-mails did not 'bounce', they were simply ignored. In all such cases, follow-up e-mails were sent after several weeks had gone by without any response. If there was still no response to the follow-up, the rental station was excluded from the book.

A second anomaly is that a small number of rental stations, which again appear on their owner's or other websites as being offered for rent (some famously so) are in fact no longer available. In some cases the owners have stopped letting them because they have taken up full-time residence in what was formerly a holiday home, in other cases the properties have been sold, or are on the for sale market.

Finally, in a small number of cases, the owners did not wish their property to appear in the book, even if the station is still available for rent by friends or by radio amateurs already known to the owner. Obviously, their wishes are respected and these stations have also been excluded from this book.

In order to prevent owners from receiving unwelcome enquiries and so that potential renters do not waste their time, here is a list of former or current stations that fall into one or more of the above categories and therefore can be assumed are no longer normally available for rent:

ANTIGUA: Royal Antiguan Amateur Radio Club, V26DX (station at Royal Antiguan Hotel); **ARUBA:** P40V/P49V/ AI6V villa; **ARUBA:** P40A station at *Iguana Villa*; **ARUBA:** P40L station; **AUSTRALIA:** Rick Rodgers, VK4HF, B&B near Brisbane, Queensland; **AUSTRIA:** Hotel Gallspacherhof, OE5XGN; **BRUNEI:** Tungku Lodge (V85SS station); **CAYMAN ISLANDS:** ZF1A club station; **COLOMBIA:** HK1AR station, Cartagena; **COSTA RICA:** TI2HMG Ham Vacation, near Ciudad Colon; **COSTA RICA:** Casa Talamanca Apartments, Playas del Coco (TI7WGI station); **DOMINICAN REPUBLIC:** *Ferienhaus Luzia* (DX-Urlaub, DL4NCF / HI9CF and DL4NYL / HI9NY station); **EAST MALAYSIA:** 9M8CC station, Sarawak; **EAST MALAYSIA:** Hillview Gardens Amateur Radio Club, Keningau, Sabah (9M6AAC station facilities currently unavailable; *may* re-open in future); **ENGLAND:** Brimham Lodge Farm B&B / Holiday Cottage; **GALAPAGOS ISLANDS:** *Hogar de Don Guido*, San Cristobal Island (Guido Rosillo, HC8GR, station); **GAMBIA:** Noz Bryan, C5DXC, station; **GUERNSEY:** Guernsey Amateur Radio Society (GARS) Club Station, GU3HFN (unavailable for rental in 2008 / 2009 but *may* become available again in the future); **GUAM:** Palm Tree DX Club, WH2DX / KH2JU station; **JUAN FERNANDEZ:** Eliazar Pizarro, CE0ZIS, station; **NEW ZEALAND:** *Huntaway* B&B, Rotorua (Graeme Hunt, ZL1ANH); **NORTH COOK ISLANDS**: E51WL station, Penrhyn Island; **SPAIN:** Casa DX, Torrevieja, Costa Blanca; **TURKS & CAICOS:** *Sea View Villa*, Providenciales Island; **USA:** *Lost*

Creek Cabin, Colorado; **US VIRGIN ISLANDS:** *Windwood Villa*, WP2Z, St Croix (note that the WP2Z callsign may however be used from other locations in the US Virgin Islands); **VIETNAM:** DX Shack, the 'Cherry Room', Kimdo Hotel, Ho Chi Minh City (Saigon).

Diamond Ridge Cottage, Homer, Alaska.

RENTAL STATIONS AROUND THE WORLD

The following listing is of those stations that *are* available for rent, as of mid-2008! If you are aware of any stations that are not included in this list (and have not been excluded for the reasons described on the previous page), please contact the editor at teleniuslowe@gmail.com with details.

ALASKA

Diamond Ridge Cottage, Homer
Website:
www.diamondridgecottage.com
Diamond Ridge Cottage is a new, three bedroom, two bathroom, 1600 square foot home located five miles north of Homer, Alaska. Two bedrooms each have a queen-size bed, while the third has a bunk with double and twin beds.

The third bedroom is also a fully equipped shack for use by visiting radio amateurs. The rig currently is a Kenwood TS-430S plus a Heathkit SB-200 linear amplifier. Antennas include a Hy-gain TH-7DX at 60ft, a full-size 80m ground plane with elevated radials, the top at 120ft, plus various dipoles and a Hustler 5-BTV five-band vertical.

You are welcome to bring your own rig along to operate if preferred and Dave, AL7DB, can help with interface cables to connect your rig to the linear or antennas. High-speed DSL and 2.4GHz Wireless Internet are available.

Dave and his wife Eileen own three VHF / FM commercial radio stations and one on 620kHz AM. Two stations, KWVV-FM 103.5 (100kW) and KGTL AM 620 (5kW) are located next to their home and the rental cottage. Operation on all bands is no problem with only two frequency exceptions: 3720 and 14260kHz (both harmonics of 620kHz). KGTL's 400ft tower is about 300ft away so it is difficult not to be affected. However, the rest of the amateur band spectrum is clear with no significant interference or overload issues.

The cottage has a TV, VCR, DVD, washer and dryer. Local phone calls are free of charge. Outside, there is a large wooden deck with sweeping views of Cook Inlet and Mt Augustine volcano. Fishing and hunting are popular activities in the local area. Homer is located roughly 220 miles from Anchorage's Ted Stevens International Airport and

The rental shack at Diamond Ridge Cottage.

Dave Becker, AL7DB, on the air.

can be driven in four to five hours.

Diamond Ridge Cottage is available for daily, weekly or monthly self-catering rental. No food or meals are offered as part of the rental agreement, although coffee, tea and cocoa are provided. The cost is USD $200 per night for one or two people, including use of the ham shack by licensed radio amateurs, plus $25 extra for each additional person.

Contact: Dave (AL7DB) and Eileen Becker, PO Box 109, Homer, Alaska (AK) 99603, USA.

Tel: +1 907 235 7526.

E-mail: al7db@arrl.net *or* info@diamondridgecottage.com

ANTIGUA

Team Antigua, V26B station (IOTA NA-100)
Website:
www.n3oc.dyndns.org/v26b

The V26B station is available for rent. It is usually booked for the four ARRL DX and CQWW DX contest weekends but is often available for both the *CQ WPX* contests, the RSGB IOTA contest, the ARRL 10m contest, all the 160 contests, WAE etc.

The station is well-equipped with antennas, but the shack itself is fairly basic and many DXpeditioners elect to stay elsewhere and use the shack only for operating. At the shack are two beds, a small cold water shower and a small kitchen with refrigerator and propane stove. There is no hot water and soap, toilet paper, towels and rubbish bags are not provided.

There are seven towers with the following antennas. 10m: 5/5/5 stack, 15m: 5/5 stack; 20m: 3/3 stack; 40m: 2/2 'shorty-forty' stack; 80m: two fixed

The shack and 40m stacked Yagis at V26B.

wire beams, a 3-element towards Europe (between the 15 and 20m towers) and a 2-element for USA (between the 15 and 40m towers); 160m inverted-Vee at the top of the 10m tower. There is a field behind the shack that was cleared in October 2007 and US and EU Beverage receive antennas laid out over the brush.

Operators normally take their own equipment, but it may be possible to borrow an HF transceiver and a Dentron MLA-2500 linear amplifier. Please en-

"...the shack itself is fairly basic..."

quire for details.
Contact: Sam Harner, WT3Q, 893 Narvon Rd, Narvon, PA 17555, USA.
E-mail: wt3q@comcast.net
or Roy Carty, V21N/W4 (station owner, living in Miami):
Tel: +1 954 894 7602;
mob: +1 954 801 6061.

AUSTRALIA

VK3QB holiday home, Venus Bay, Victoria (IOTA OC-001)
Website:
www.vk3qb.com/index2.htm
(follow the 'dx holiday' link)
Venus Bay is a small seaside village approximately 160km (about 90 minutes by car) south-east of Melbourne. The property is located five minutes walk to the shops and a restaurant and about 10 minutes to the beach. The area has some beautiful ocean beaches, good fishing and many miles of walking tracks through the national park. Five minutes up the road is a pub with cheap beer, good food, pin ball machines and a band at the weekend during the summer.

The property has three bedrooms, with double beds and bunks for four people. The radio equipment includes Yaesu FT-920 and FT-101ZD transceivers and an FL-7000 solid-state linear amplifier to a 3-element triband beam and an 80m dipole at 35ft. Also in the shack is a 1kW roller inductor MFJ ATU, an electronic keyer and Kent paddle key.

For more information about this property, availability and rates please contact the owner, Chris, VK3QB.
Contact: Chris Chapman, VK3QB, 14 Saturn Parade, Venus Bay, Victoria 3146, Australia.
E-mail: vk3qb@hotmail.com

VK3QB Holiday Home, Venus Bay.

BAHAMAS

Sand Dollar Villa, Treasure Cay, Great Abaco (IOTA NA-080)
Website: www.sanddollarvilla.com/ hamradio.htm
Sand Dollar Villa is located at Treasure Cay on the island of Great Abaco in the Bahamas. Abaco is located 175 miles due east of West Palm Beach, Florida. Treasure Cay has its own airport eight miles from the resort. Commercial and charter flights are available from Miami, Fort Lauderdale, West Palm Beach and Orlando, as well as Nassau, Bahamas.

Sand Dollar Villa sleeps four in two bedrooms with two private bathrooms, indoor and outdoor tiled patios, and a fully-equipped kitchen.

Although the owners of Sand Dollar Villa are not ham radio operators themselves, the villa is equipped with vertical antennas, power supplies, ATU and a tool kit. The only thing a guest needs to do is bring their own radio. The antennas are a Butternut HF6V vertical for 80, 40, 30, 20, 15 and 10m and a Cushcraft R6000 vertical for 20, 17, 15, 12, 10 and 6m. The villa is equipped with a Samlex SEC-1223 power supply, with an Astron RS-20 power supply as back-up. There is an MFJ-941C antenna tuner for 80 - 10m with coax jumpers to the radio and a built-in antenna switch. The villa has its own wireless hotspot for Internet access.

A 3.5-mile long white sand beach, described by *National Geographic* as "one of the 10 most beautiful beaches in the world", is a short walk from the villa,

Sand Dollar Villa, Treasure Cay, Great Abaco, Bahamas.

as are an 18-hole golf course, a full-service marina, tennis courts, several restaurants and beach bars, boat rentals and a professional dive shop. World-class bonefishing and sport fishing with expert guides, snorkelling and scuba diving are all available. The villas' swimming pool offers direct access straight to the beach.

Contact: Matt Smith.
E-mail: owner@sanddollarvilla.com

BELIZE

The Shack at Palmetto Place Guesthouse, Placencia
Website: www.wishwilly.net
The Shack is located adjacent to (100 feet from) the Palmetto Place Guesthouse in Placencia, southern Belize, and is available for rent on a daily or weekly basis, whether or not you are staying at the Guesthouse.

The Shack is in a comfortable, stand-alone building with ceiling fan, full-size couch and three large screened windows. It is equipped with a Kenwood TS-850S transceiver with narrow CW and SSB filters, an Ameritron ALS-600 600W amplifier, headset, foot switch, clear speech speaker, paddle and straight key, and RS232 computer output for your laptop. A Kenwood TS-690 is dedicated to 6m but is also available as a back-up HF transceiver. The antenna line-up comprises a five-band Lightning Bolt cubical quad on a 10m tower, a Butternut HF2 vertical located 70ft out in the salt water lagoon, a Butternut HF9V vertical for 80 to 6m, and a 4-element wide-spaced 6m beam on its own rotator (all equipment is subject to change and upgrading).

Rental of the Shack costs USD $60 per day or $350 per week ($300 per week if you are staying at the Guesthouse). To book the Shack, contact Bob Fox, V31MD (see below). The Guesthouse is available from $500 per week for one or two people and reservations are handled exclusively by Caribbean Tours of Placencia (see www.ctbelize.com).

Contact: Bob Fox, V31MD, PO Box 73, Placencia, Belize.
Tel: +501 523 3425.
E-mail: bob@wishwilly.net

BERMUDA

Tarrafal Apartments (No 5 and No 7), Hamilton Parish (IOTA NA-005)
Website: http://vp9ge.com
Tarrafal Apartments Nos 5 and 7 are available for your holiday operation in Bermuda. The apartments are located about 20 minutes from the City of Hamilton and buses stop at the gate every 15 minutes. The apartments are designed for singles or couples and can take one extra person on a convertible couch. No 5 is a large bed-sitter and No 7 is a one-bedroom apartment. Each unit has air-conditioning, providing both heating and cooling, bathroom and kitchen with stove, fridge, microwave, toaster, coffee maker, pots and pans, etc. Please note that these are both no-smoking apartments, although smoking outside on the porch is OK.

The antennas available include a Cushcraft A4S 4-element Yagi for 10, 15 and 20m on a Strumech tower with rotator, a Cushcraft R-6000 vertical, a G5RV for 6m to 160m, a 160m inverted-L, an excellent dipole for 40m, and Yagis for 6m and 2m. You should take your own computer for logging and if possible also take your own rig (13.8V power supplies available). It may be possible to borrow a rig, but check with Ed first to avoid disappointment. The power limit in Bermuda is 150W input to the finals so please do *not* take a linear amplifier to the island. ADSL broadband Internet is available.

Membership of the VP9I DX Club

The Cushcraft A4S beam and 40m rotatable dipole at Tarrafal Apartments.

is open to all radio amateurs and this callsign may be used by club members during contests to help overcome any problems encountered by signing "homecall/VP9".

Close to the apartments are an English-style pub, grocery, post office, the Crystal Cave, and ice cream factory and snack bar, as well as golf, tennis and water sports facilities.

Free transportation from and to Bermuda International Airport and to a nearby beach is provided and every seventh night's accommodation is free for radio amateurs. The rates depend on the length of stay and the number of persons but in 2007 were USD $150 per night per couple. However, with fuel and other costs having increased there may need to be an adjustment soon. All taxes and service charges are included. **Contact** (use airmail *only*): Ed Kelly, VP9GE, PO Box 1555, Hamilton HM-FX, Bermuda.
Tel: +1 441 293 2525;
fax: +1 441 295 3559.
E-mail: ed@vp9ge.com

BHUTAN

Bhutan Ham Centre
Website: www.proteustours.com.bt/ham/index.html
Proteus Tours and Travels, owned by Yeshey Dorji, A51AA, operates the Bhutan Ham Centre and can arrange licensing, accommodation and a well-equipped station for the visitor to Bhutan.

Tower and triband beam at Bhutan Ham Centre.

Multiband vertical, Bhutan Ham Centre.

The station has no fewer than eight transceivers: Icom IC-756 ProII, Icom IC-746, Kenwood TS-140S, Kenwood TS-680S, Kenwood TS-440S, Yaesu FT-900, and 2 x Yaesu FT-747GX; and five linear amplifiers: Yaesu FL-2100Z, Yaesu FL-2100B, Tokyo Hy-Power HL-66V and 2 x Dentron MLA-2500. There are seven beam antennas mounted on two 10m towers, plus verticals, as follows: Create HF Yagi, Mosley TA-33M Yagi for 10, 15, 20m, 3 x Cushcraft A3S Yagis for 10, 15 and 20m, Cushcraft A3WS 3-element Yagi for 12 and 17m, Create 730V multi-band V-dipole, Cushcraft R7000 vertical, Cushcraft R8 vertical, Gladiator 160 / 80m vertical, 6m Yagi. In addition there are band-pass filters for each band to allow for 'multi-multi' type operation, and ATU, power meter, rotors etc.

Use of the Bhutan Ham Centre's radio equipment is free of charge but all visitors to Bhutan must pay the obligatory daily 'Minimum Tourist Tariff'. The tariff is fixed by the Bhutan government's Department of Tourism (DoT). All tour operators in the Kingdom should charge no more or no less than the rates fixed by the DoT. In the peak season (March, April, May, September, October, November) the tariff is USD $200 per person per night, while in the off season (December, January, February, June, July, August) it is $165. These rates are for groups of three or more. There is a surcharge of $40 per night for an individual or $30 per person per night for a group of two. The minimum daily

Bhutan Ham Centre: 80m vertical overlooking Thimphu Dzong.

rate includes accommodation, food and beverages (but not alcohol), services of guides, transport to any place within Bhutan, riding ponies and pack animals on treks, trekking staff and equipment, and a royalty payable by all tour operators to the DoT. The rates apply uniformly, irrespective of location and type of accommodation provided.
Contact: Yeshey Dorji, A51AA, PO Box 73, Thimphu, Bhutan.
Tel: +975 2 323242 / 325333 / 325353; **fax:** +975 2 325888.
E-mail: proteustours@druknet.bt

BOTSWANA

African DX Safaris
African DX Safaris can organise tailor-made DXpeditions to Botswana, among other countries, providing accommodation, meals, equipment and antennas. See main listing under SOUTH AFRICA.

CAMBODIA

DX Shack, Sihanoukville
Website: www.qth.com/dxshack/ XU/XUtop.htm
The DX Shack in Sihanoukville (also known as Kompong Som) was set up by Hiroo, XU7AAA (ex-JA2EZD), at the Hill

Side View Guest House, a small private guest house run by a friendly Cambodian family. It is located in the Victory Hill area of the city, on top of a small hill, with a view to the sea.

The guest house offers basic accommodation. The DX Shack is in room number 202 and has a king-size bed, air conditioning unit, and a bathroom with running cold water, wash basin and toilet (flushed by buckets of water).

The equipment provided is a Kenwood TS-690 HF / 6m 100W transceiver and Tokyo Hy-Power HL-1Kgx linear amplifiers. The antennas are a Create Design CD-318 Jr 4-element Yagi

The Hill Side View Guest House at Victory Hill, Sihanoukville, Cambodia, host for the Cambodia DX Shack.

PHOTO: 9M6DXX

PHOTO: 9M6DXX

Create Design HF Yagi and low-band dipoles at Sihanoukville DX Shack, Cambodia.

for 10, 15 and 20m, as well as inverted-Vee dipoles for 40, 80 and 160m on a 22m high tower.

Sihanoukville is the second largest city in Cambodia and is 250km south of the capital, Phnom Penh. These days there is a good road from Phnom Penh all the way to the town. Buses make the journey five times a day in each direction and cost around USD $6, while a taxi from the city centre or airport will cost from $60 to $100.

Although Cambodia has a reputation for being a dangerous place, these days it is generally as safe as many Western countries and Sihanoukville in particular has a laid-back feel. The centre is a bustling modern city, but the Victory Hill area has many small hotels, restaurants, bars, discos and other facilities for tourists. There are several sandy beaches close to the town and trips to the offshore islands, which count as AS-133 for IOTA, can be arranged (note though that the Cambodian amateur radio licence is for a single specific location only: if you wish to operate from the offshore islands you will need an additional licence with separate callsign).

The basic rental fee (for accommodation and use of the transceiver and antennas) is $38 per night. Room 202 has its own electricity meter and in addition you pay the owners of the guest house $0.90 per kWh of electricity consumed by the air conditioning unit, amplifier (if used), rig etc.

Reports in 2008 are that the equipment and wire antennas are in a poor state of repair and the air-conditioning

unit in Room 202 is no longer working. If this is of concern to you, check the current state of the DX Shack with Hiroo before booking.

Contact: Hiroo Yonezuka, XU7AAA/HS.
Mob: +66 81 778 713.
E-mail: dxshackmgr@hotmail.com

CANADA

PEI DX Lodge, Prince Edward Island (IOTA NA-029)
Website: www.peidxlodge.com
The PEI DX Lodge is a true 'super-station' available for rent. It is a 3000 sq ft farmhouse on 1.67 acres at Bloomfield, about 75 to 90 minutes from Charlottetown airport. The farmhouse was bought by Ken Widelitz, K6LA / VY2TT, in 2002 and he has transformed it into his dream QTH. Ken uses PEI DX Lodge for five or six major contests per year: the rest of the time it is available for rent.

There are four operating positions in the custom-designed 17 x 13ft shack which can be configured for single operator, SO2R, multi-single, multi-2 or multi-multi use. The transceivers are an Icom IC-7800, Icom IC-756 ProIII, 2 x Icom IC-756 ProII and a Kenwood

The PEI DX Lodge on Prince Edward Island, Canada, from the front.

Looking from the 20m antenna towards the 40m tower.

Just some of the PEI DX Lodge antennas.

TM-261 for 2m. Amplifiers: 2 x Alpha 87A, Alpha 91b, Dentron MLA-2500.

Outside, there are no fewer than five towers with a 160m dipole at 130ft, a Force 12 EF180 160m vertical with elevated radials, an 80m 4-Square array with elevated radials, a full-size 3-element M^2 40m beam at 145ft, a 2-element 402BA at 95ft (fixed), a 6-over-6-over-4 array for 20m on a 150ft tower, a 6-over-6-over-4 array for 15m on a 140ft tower, a 7-over-7-over-4 array for 10m on an 84ft tower, as well as a TH6DXX tribander and antennas for 2, 6, 12 and 17 metres. All rotatable monoband antennas for 40m and above are fed with 7/8in hardline and custom-built antenna switching by WX0B allows any of 12 antennas to be selected at any of four operating positions. Ken points out, *"The antenna and tower photos [on the website] are not completely up to date and the description changes as improvements are made."*

The lodge can sleep seven in four bedrooms and has three bathrooms. There is a new private lounge with satellite TV, DVD and VCR adjacent to the shack and a kitchen with washer/dryer. As a break from the radio, or for other members of the family, a resort with 18-hole golf course is less than a mile away, with a water park about five miles distant.

The standard PEI DX Lodge rental period is from Wednesday to Monday and rates are based on the number of amateur radio operators: there is *no charge* for non-licensed guests. The first operator pays USD $1000, the second operator $500 and each additional operator $250. Additional days are charged at $125 per operator, if there are no guests booked for the next rental period.

Contact: Ken Widelitz, K6LA / VY2TT, 10519 Lauriston Ave, Los Angeles, CA 90064, USA.
E-mail: widelitz@gte.net

CANARY ISLANDS

El Porche, Telde, Gran Canaria (IOTA AF-004)
Website: www.qsl.net/ea8azc
El Porche is located in the north-east of Gran Canaria, 6km from Telde and 15km from Las Palmas. Adjacent to the

The PEI DX Lodge operating shack.

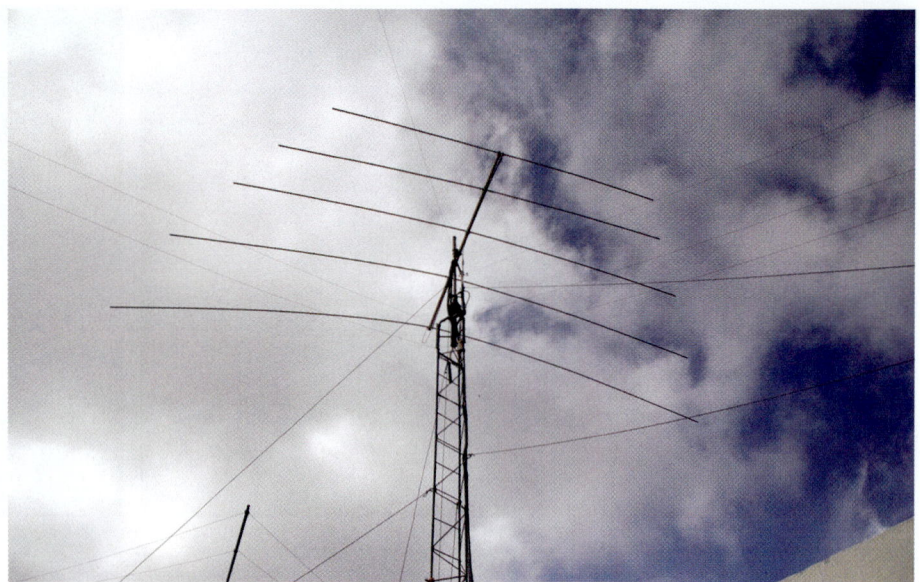

The Titanex LP-5 log-periodic beam at El Porche, Canary Islands.

main home is a bungalow in a court-yard with flowering shrubs, a fish pond and palm trees which is offered to paying guests seeking comfort and tranquillity in a completely private atmosphere.

The guest accommodation features a living room with English and German satellite TV channels, fully-fitted kitchen with coffee maker, toaster and microwave oven, two bedrooms with large double beds, fitness room, solar-heated swimming pool and a sun terrace with views towards Telde and the coast.

The amateur radio equipment available includes a Kenwood TS-570D and an Icom IC-706 transceiver, a restored Collins S-Line and a computer loaded with various amateur software. Anten-

nas available are a Titanex LP-5 log-periodic beam, a home-made Spiderbeam, a Barker & Williamson wide-band dipole for 3 - 30MHz and a trap dipole for 40, 80 and 160m. Broadband ADSL Internet access is available on request.

Telde is the oldest community in the Canary Islands and a visit to this historic city with its traditional plazas, churches and ancient buildings is highly recommended. Lately Telde has grown into a modern and bustling community with supermarkets and other shopping facilities. Regular concerts and exhibitions add to the unique flavour of the town. Spain's oldest golf course, where foreign guests are welcome, is close by. The nearest beach is

The radio shack at El Porche, Gran Canaria.

Home-made Spiderbeam (left) with Titanex log-periodic.

The Christopher Columbus Condominiums on Seven Mile Beach.

6km away, while the overcrowded tourist enclave of Playa del Inglés is 45km to the south - and a world away.

The *el Porche* bungalow is suitable for a maximum of two adults and two children up to 13 years of age. Please contact Bill, EA8AZC, or Brigitte Kiwitt to ask for the price. Bill says: *"The el Porche estate is on the market for sale on a long-term basis. This will certainly take years considering the slack market conditions as well as the asking price. Meanwhile paying guests are still very welcome and we will continue to enjoy the place for years to come."*

Contact: Bill (EA8AZC) and Brigitte Kiwitt, *el Porche*, Palmital - Alto No 17, 35200 Telde, Gran Canaria, Spain.
Tel: +34 928 572410.
E-mail: bbkiwitt@gmail.com

CAYMAN ISLANDS

Unit 24, Christopher Columbus Condominiums, Seven Mile Beach, Grand Cayman (IOTA NA-016)
Website: www.martykaiser.com/ 24.htm *and*
www.christophercolumbuscondos.com
The Christopher Columbus Condominiums, located right on Grand Cayman's

Unit 24 at the Christopher Columbus Condominiums.

famous Seven Mile Beach, have two bedrooms and two bathrooms with a living room, dining area, kitchen and large sliding glass doors that open on to a patio.

Unit 24, just 250 feet from the Caribbean Sea, comes with a Cushcraft R6 multi-band vertical (6 - 20m) and a 106ft long-wire antenna and is available for rental by licensed radio amateurs. You should bring your own rig and antenna tuner.

There are reduced rates available for radio amateurs if you book through the Unit 24 website. See the website for booking instructions. Daily rate charges for up to four persons: Low season (mid-April to mid-December) USD $160.00; High season (January to mid-April) $290.00; Peak (mid-December to New Year) $340.00. Each additional person is $25.00. 6% gratuity charged in lieu of tipping, plus 10% Government Tax.
Contact: Marty Kaiser, PO Box 171, Cockeysville, MD 21030, USA.
Tel: +1 410 252 8810.
E-mail: martykaiser@prodigy.net

COOK ISLANDS

See South Cook Islands.

COSTA RICA

TI5KD / TI5N station, San José
Website: www.yantis.us/ti5kd
Carlos ('Keko') Diez, TI5KD, and his wife Sophia offer accommodation at their home in the semi-rural suburb of Alajuela. Either all meals or no meals are provided, depending on the number

PHOTO: W4GKF

Sunset over the TI5KD antennas.

Your host Keko, TI5KD.

of guests. The accommodation comprises an 'efficiency' (self-catering) apartment and shack plus another complete efficiency apartment without shack that will sleep four. The maximum number of guests is eight.

With several high towers supporting monoband and five-band HF quad and Yagi beams, the location is excellent for operating. Keko points out that if you are bringing your family who hope to enjoy a swimming pool, room service and a spa you would be better off staying at the nearby five-star Marriott hotel. Most ham visitors, however, are happy just to use the radio, enjoy Sophia's great cooking and have a bed to 'crash' on when the pile-ups allow!

The equipment consists of two Kenwood TS-940s and several amplifiers including an Alpha 89 and some Henrys. Guests need to take their own logging laptop computer and may also take their own rigs if they prefer. There are two or more antennas for every band from 160m to 70cm, including a quad for 10, 12, 15, 17 and 20m, and a 2-element quad for 40m (fixed on north).

An Internet connection is available 24 hours a day and there is a telephone in the shack. The TI5N special contest call is available for use from Keko's station. A Costa Rican permit is required in order to operate and you should send copies of your licence and passport in advance.

Keko's home is about 15 minutes from San José airport, about 10 minutes from the Marriott and 30 minutes from the city centre. Although San José is not really a tourist destination, Costa Rica most certainly is, with jungles, volcanoes and many miles of beautiful beaches to explore. Keko suggests combining an operating stint from his station with either a stay at one of the fancy resorts on the Pacific coast or 'roughing it' in the jungle.

The rates vary with the number of guests, whether or not it is a contest period and if meals are to be included or not. Interested parties should contact Keko by e-mail. Although he is not in the tour guide business he will do tours if he has time and if it is what you want to do - please ask.
Contact: Keko Diez, TI5KD, PO Box 195-4005, Belen, Heredia, Costa Rica.
E-mail: ti5kd@yantis.us

CRETE

**Pelamare Apartments, Kokkini Hani, nr Iraklion (IOTA EU-015)
Website:
www.freewebs.com/pelamare**
Pelamare is a complex of 30 fully furnished apartments in an attractive garden by the beach. Located 13km east of Iraklion in the village of Kokkini Hani, Crete, Pelamare Apartments each house two to four people. They include com-

PHOTO: N4NX

The TI5KD radio shack.

Pelamare Apartments, Crete.

plete kitchen facilities, refrigerator, bathroom, bedroom, direct dial telephone, music and hairdryer. TV with internal satellite and video system can be made available on request as can the use of a safe. Other facilities available in the complex are a swimming pool, lounge with TV, snack bar, table-tennis and sauna. A sandy beach is nearby. There is a wide variety of restaurants and supermarkets within walking distance.

Use of the shack is free to all radio amateurs who stay at Pelamare Apartments, but they must take their own transceiver. In the shack there is a power supply and the antenna is a Windom wire antenna.

Contact: Manos Nerantzulis, SV9ANJ, Pelamare Apartments, PO Box 1272, Iraklion, Crete 71110, Greece.
Tel: +30 2810762000;
fax: +30 2810762000
E-mail: pelamare@her.forthnet.gr

Pool at Pelamare Apartments.

CROATIA

9A3MR station, Jezera, Murter Island (not IOTA)
Website: www.qsl.net/9a3mr

The 9A3MR contest QTH on Murter Island is available for rent during the summer season. There are two apartments available: the first is a four-room apartment on the first floor with a large dining room and kitchen, while the second is a small studio apartment suitable for two people and one child. The house is located 200m from the beach and has its own parking place and barbecue area.

The 9A3MR station on Murter Island, Croatia.

The Optibeam Yagi at 9A3MR.

Jezera on Murter Island, showing the location of the 9A3MR station.

The transceiver is a Kenwood TS-930 and antennas are an Optibeam OB7-3 7-element beam for 10, 15 and 20m and a 2m beam on a 16m tower.

The sea is very clean and if you like scuba diving, Murter Island is the place for you. The Kornati islands national park (IOTA EU-170) is 10 nautical miles away and tourist tours to Kornati are organised daily. Krka national park, approximately 25km away, can also be reached by car. It is possible to organise full-day sailing trips on the owner's sailing boat.

The price for radio amateurs is EUR 32 euros (approx GBP £23) per night in the high season (July and August) and 25 euros (approx £18) the rest of the year for the small apartment, and 90 euros (approx £65) in the high season, or 70 euros (approx £50) in the low season for the big apartment:

Contact: Rolando Milin, 9A3MR, Put Ravnih Njiva 30, Split 21000, Croatia. **E-mail:** rolando.milin@st.htnet.hr

DOMINICA

J73HPL cottage, Wotton Waven (IOTA NA-101)
Websites:
http://va-mountainland.com/ DomCottage/DomCabin.html *and* **www.j7hams.com/j73cpl_hpl.htm**

A cottage located in the village of Wotton Waven in the cool mountains about four miles from the capital Roseau is available for rent. The elevation and lack of power line or industrial noise makes this an ideal radio location.

The amateur radio station is well equipped with 500W on 160 - 10m, a Mosley TA-33 3-element beam with 17 and 12m kit covering all five bands between10 and 20m on a 38ft tower, a Carolina Windom for 160 - 30m and vertical antennas for VHF / UHF. The cottage is located on a ridge approximately 800ft up above the capital city of Roseau.

The cottage can accommodate singles, doubles, and groups of up to five people. The price is USD $50 per night for one or two people or $120 for the whole cottage, whether it's three hams or a family of husband, wife and three children. Stay seven nights and pay just for six: it's $720 for a whole week.

The J73HPL Cottage is recommended by Dave (KK4WW / J79WW) and Gaynell (KK4WWW / J79WWW) Larsen, who provide the website for Hetty and Clement and who may also be able to provide further information.

Contact: Hetty (J73HPL) and Clement (J73CPL) Pierre-Louis, Wotton Waven, Dominica, West Indies.
Tel: +1 767 449 9692.
E-mail: j73hpl@hotmail.com
or

The J73HPL Cottage on Dominica.

View from the tower at J73HPL.

Mosley 5-band beam and Carolina Windom at the J73HPL station.

J73HPL Cottage contact details cont:
Contact: Dave (KK4WW) and Gaynell (KK4WWW) Larsen, PO Box 341, Floyd, VA 24091, USA.
Tel: +1 540 745 2322, *or*
+1 540 763 2321.
E-mail: land@swva.net *or*
dlarsen@swva.net

EAST MALAYSIA

EAST MALAYSIA – SABAH (9M6)
Eagle Plateau, nr Keningau
(IOTA OC-088)
Website: www.eagleplateau.co.nr

Eagle Plateau is a 50-acre ecological project and the home of Alfons Undan, 9M6MU, and his wife Doris, 9M6DU. At 2500ft ASL, Eagle Plateau is located above the town of Keningau in the interior of Sabah, some 2.5 hours by car from the international airport at Kota Kinabalu. There are spectacular views of Mt Kinabalu (4095m) and Mt Trus Madi (2642m), the highest and second highest peaks in Malaysia, as well as

PHOTO: 9M6DXX

Alfons, 9M6MU, at Eagle Plateau.

PHOTO: 9M6DXX

the backbone of Sabah, the Crocker Range. Because there is little or no light pollution, at night Eagle Plateau becomes a natural planetarium!

Eagle Plateau was created as a family sanctuary using clean renewable energy technology including photovoltaic, solar thermal, wind power and micro hydro-electric systems. There is no mains electricity on site so power is generated using solar panels charging banks of batteries. Their own water supply is pumped using the latest solar pumping technology.

Alfons says, "*We are still busy building a lot of the infrastructure at Eagle Plateau but it is slowly taking shape and we are excited about the whole idea of creating a place that will be truly relevant to the enjoyment of amateur radio at its fullest. When fully completed, it will be a modest station with all the stuff we hope all amateur radio enthusiasts will enjoy.*"

At present the amateur radio station offers low-power operation (due to the restrictions of the renewable energy source) with a tower, 3-element triband beam and dipoles. It is an excellent place to relax, enjoy amateur radio operating and to see for yourself just what can be achieved using only natural resources and renewable energy technology.
Contact: Alfons (9M6MU) and Doris (9M6DU) Undan, PO Box 210, 89008 Keningau, Sabah, Malaysia.
Tel: +60 87 338 500;
mob: +60 16 914 9383.
E-mail: alfons5231@gmail.com

Guest accommodation at eco-friendly Eagle Plateau.

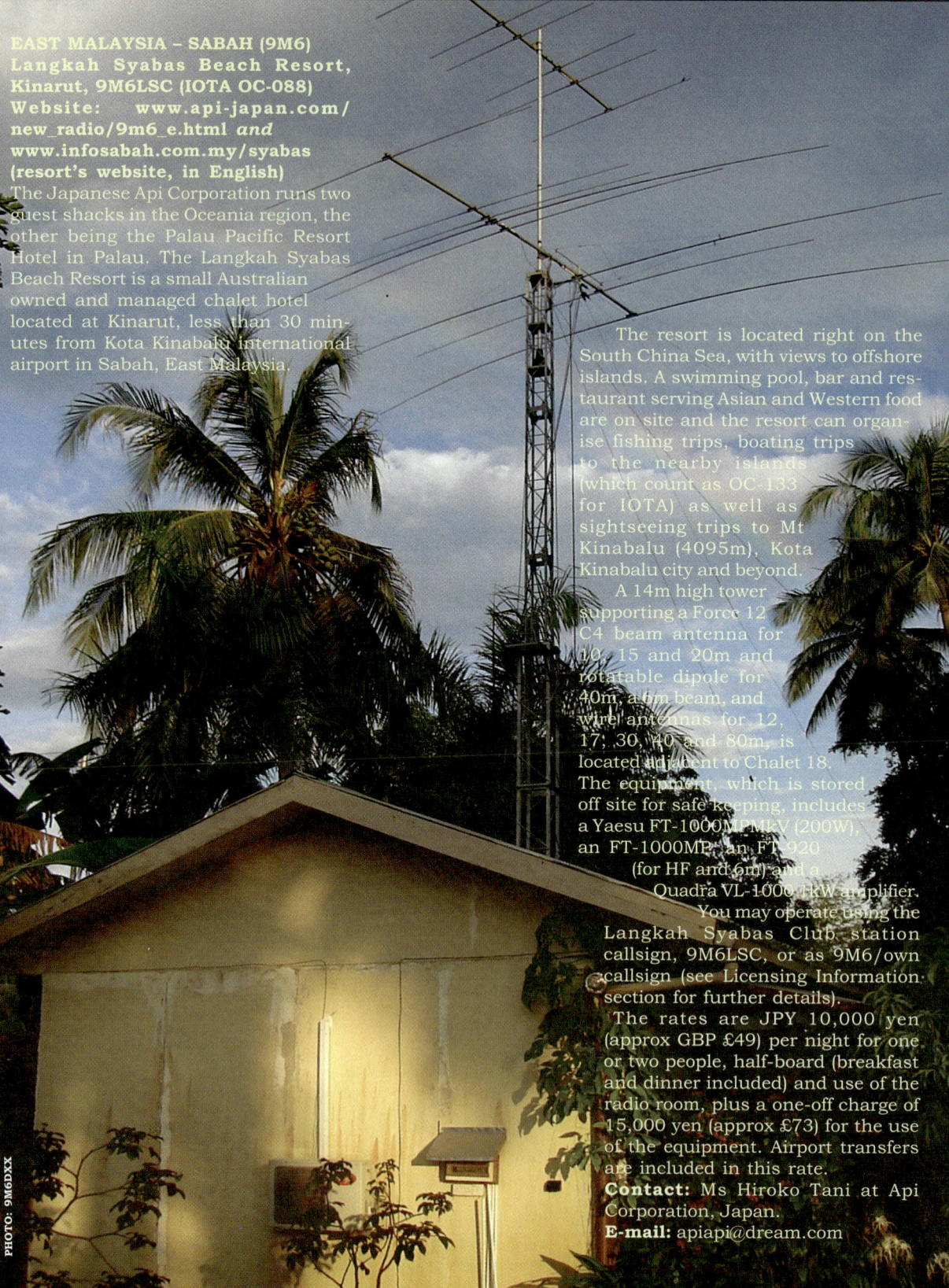

EAST MALAYSIA – SABAH (9M6)
Langkah Syabas Beach Resort, Kinarut, 9M6LSC (IOTA OC-088)
Website: www.api-japan.com/new_radio/9m6_e.html *and* www.infosabah.com.my/syabas (resort's website, in English)

The Japanese Api Corporation runs two guest shacks in the Oceania region, the other being the Palau Pacific Resort Hotel in Palau. The Langkah Syabas Beach Resort is a small Australian owned and managed chalet hotel located at Kinarut, less than 30 minutes from Kota Kinabalu international airport in Sabah, East Malaysia.

The resort is located right on the South China Sea, with views to offshore islands. A swimming pool, bar and restaurant serving Asian and Western food are on site and the resort can organise fishing trips, boating trips to the nearby islands (which count as OC-133 for IOTA) as well as sightseeing trips to Mt Kinabalu (4095m), Kota Kinabalu city and beyond.

A 14m high tower supporting a Force 12 C4 beam antenna for 10, 15 and 20m and rotatable dipole for 40m, a 6m beam, and wire antennas for 12, 17, 30, 40 and 80m, is located adjacent to Chalet 18. The equipment, which is stored off site for safe keeping, includes a Yaesu FT-1000MPMkV (200W), an FT-1000MP, an FT-920 (for HF and 6m) and a Quadra VL-1000 1kW amplifier. You may operate using the Langkah Syabas Club station callsign, 9M6LSC, or as 9M6/own callsign (see Licensing Information section for further details).

The rates are JPY 10,000 yen (approx GBP £49) per night for one or two people, half-board (breakfast and dinner included) and use of the radio room, plus a one-off charge of 15,000 yen (approx £73) for the use of the equipment. Airport transfers are included in this rate.
Contact: Ms Hiroko Tani at Api Corporation, Japan.
E-mail: apiapi@dream.com

PHOTO: 9M6DXX

EAST MALAYSIA - SARAWAK (9M8)
The Holiday Inn, Kuching
(IOTA OC-088)
Website: www.holidayinn.com

A station belonging to the Sarawak Amateur Radio Club is situated on the top floor of the Holiday Inn hotel in the state capital, Kuching. A Kenwood TS-140S and Hy-Gain 'Exporer-14' 4-element Yagi for 10, 15 and 20m are available to club members and to any amateurs staying at the hotel.

All radio amateurs with a valid Malaysian licence can operate from the station free of charge. The station is located in a utility room adjacent to the lift motor room on the 11th floor of the hotel. The key to the radio room can be obtained by making arrangements with Allan Ming, 9M8MA (not the hotel). Although the Sarawak Amateur Radio Club makes no monetary charge for the use of the station, donations such as good quality connectors (e.g. N-type or PL-259s) or even simple test gear would be appreciated.

Hotel reservations must be made with the Holiday Inn Kuching: the hotel does not offer any special rates for radio amateurs, but package deals are often available. The Holiday Inn is located on the Kuching waterfront with views over the Sarawak River, yet is also in the heart of the city's business, entertainment and shopping district. The hotel offers a sauna and an outdoor pool. Local attractions include historic Fort Margherita, the Tua Pek Kong Chinese temple and the city's main bazaar, all just 1km away, the Sarawak Museum (3km) and the Bako National Park (35km). The hotel is 20 minutes (11km) from Kuching International Airport and the taxi fare is about MYR 20 ringgit (approx £3.20) each way. Rooms rates start at about 175 ringgit (approx £28) per night.

Contact: Allan Ming, 9M8MA, Malaysian Airline System (Pre-Flight), Jalan Song Thian Chiok, 93100 Kuching, Sarawak, Malaysia.
Tel: +60 13 8164227 *or* call on 2m repeater 147.950MHz -600kHz (no tone access)
E-mail: elizas@tm.net.my *or* allanaddie@yahoo.co.uk *(for booking station)* and:
Holiday Inn Kuching, Jalan Tunku Abudul Rahman, PO Box 2928, Kuching 93756 Sarawak, Malaysia.
Tel: +60 82 423111;
fax: +60 82 426169 *(for booking hotel).*

ENGLAND

G4SOF bed & breakfast, Bideford, Devon (IOTA EU-005)

Jeff Blight, G4SOF, and his wife Teresa run a family business that includes a car dealership, fuel garage, car hire and a convenience store, as well as a bed and breakfast at the family home. Radio amateurs staying at the B&B are welcome to use Jeff's well-appointed shack for some operating while in North Devon. Jeff says, *"I'm OK with hams using the shack for casual operating though not for long contests as I use my shack every day because I work and live in the same location."*

The shack includes a Yaesu

G4SOF station antennas.

The Holiday Inn in Kuching, Sarawak.

G4SOF station in Bideford, England.

Bora Bora.

FT-2000D transceiver with DMU Data Management Unit, a Kenwood TL-922 linear amplifier, and various 2m / 70cm equipment. Antennas are a 3-element Jaybeam Yagi for 10, 15 and 20m on a 60ft Versatower, a 2-element homebrew cubical quad for 12 and 17m on a 35ft tower, and various wires and dipoles. At present Jeff also uses phased verticals for 40 and 80m, but it is likely these will have to come down when the field in which they are located is developed. The GB3ND 70cm repeater is at Jeff's location and it is on *Echolink* 24/7.

Bideford is a popular holiday location and there is plenty to do in the area. The ship for Lundy Island (IOTA EU-120) leaves from Bideford. The nearest train station is Barnstaple, 10 miles away.

Bed and breakfast accommodation is offered at around GBP £20 per person per night.

Contact: Jeffrey Blight, G4SOF, Lowbell, Handy Cross, Bideford, Devon EX39 3ET, England.
Tel: +44 1237 424011.
E-mail: jbg4sof@aol.com

FRENCH POLYNESIA

Mai Moana Island, Bora Bora (IOTA OC-067)
Websites:
www.mai-moana-island.com *and* **www.members.cox.net/ronie1/bb.htm (two different websites)**
Mai Moana Island is a private *motu* (small island) in the lagoon of Bora Bora. There are just three bungalows on the beach available for rent - or you can even rent the whole island for a very reasonable fee!

All three bungalows are located close to the waterline. They are spacious and have a bedroom with two full-size beds, TV, a dressing room, bath-room with hot water - and a perfect view of the lagoon from your verandah. The water is heated by a solar energy system and electricity provided by a generator. As there are only three bungalows there is a maximum of six visitors at a time, so it never gets crowded or noisy.

Mai Moana is owned by Stan Wisniewski, FO5IW, a Frenchman of Polish extraction, and he lives on the island with his family. His shack is available for use by licensed radio amateurs and includes a Yaesu FT-897 transceiver and an ATU and, out in the lagoon, a 3-element triband Yagi mounted on a small tower and completely surrounded by salt water. The generator is normally operational only from 6.00 - 10.00pm, but the hours can be extended for a small additional fee (there are 12V lights in the bungalows so you can have light whenever you need it). You can obtain your own FO0 licence from Papeete, Tahiti, or use the club callsign FO8DX. Stan says, *"I would appreciate it if you mention that FO8DX won first place world-wide QRP in the CQ WPX SSB contest in 1991. Also that there have been many visitors and DXpeditions*

The accommodation on Mai Moana.

Inside the bungalow.

Mai Moana.

from USA, Poland and Japan."

Breakfast, lunch and dinner are available, and you can choose accommodation only, B&B, half board or full board options. The bar in the 'round house' serves aperitifs and French wines.

There are plenty of activities to keep you occupied, including reef walking, paddling a kayak, snorkelling, or you could even be left alone for the day on one of the many uninhabited *motus* with a specially-prepared picnic lunch for two. On Bora Bora island itself are more activities such as scuba diving, a tour around the island in a *pirogue* during which you can feed sharks and meet rays, a helicopter flight above the lagoon, a yacht cruise, a jeep excursion and you can even have your own Tahitian wedding (even if you are already married!)

Daily rates are 24,000CFP (*Comptoirs Français du Pacifique Francs*) (approx GBP £145) for a bungalow for two people. An additional person is 8000CFP per night (approx £50). Radio amateurs get a 10% discount on these prices. Please note that children younger than 12 are not admitted..

Meals are extra: breakfast 1600CFP (approx £10), lunch 2500CFP (approx £15), dinner 3800CFP (approx £23) *per person.*

The half-board option (bungalow, breakfast and dinner for two people) is 34,800CFP (approx £210) per night, while full board (bungalow, breakfast, lunch, dinner for two people) is 39,800CFP (approx £240) per night.

The whole island for two people (bungalow only) is 45,000CFP (approx £275) per night, while the whole island for two (half board) is 55,800CFP (approx £340) per night. Stan, FO5IW,

or Ron, FO5VO / W6OM, can provide further details.
Contacts: Stan Wisniewski, FO5IW, Box 164, Vaitape, Bora Bora, French Polynesia.
Tel: +689 676 245 / +689 737 573.
E-mail: stan@mail.pf *or*
Ron Weaver, W6OM, 13691 Solitaire Way, Irvine CA 92620, USA.
Tel: +1 714 559 6209;
fax: +1 714 559 4832.
E-mail: w6om@cox.net

GAMBIA

4WD Portable DXpedition
Jan-François, 6W7RV, offers a portable DXpedition in his 4X4 vehicle equipped with radio from 80m to 70cm. The vehicle has two 105Ah batteries with 400/800W 12V - 220V inverter and is equipped with an SGC SG-230 SmartTuner for 1.8 - 30MHz (max 200W) with SG-202 antenna, and Maldol HFC-40L, HFC-30L and HFC-20L mobile antennas for 40, 30 and 20m. The 4X4 vehicle has a special 50 litre water tank and a 45 litre fridge. The Gambia is one of several possible destinations. For further details see the main entry under SENEGAL.

GERMANY

Hotel Hellers Krug, Holzminden
Websites: www.hellerskrug.de *and* www.dl2obo.de (two different websites)
Hotel Hellers Krug in Holzminden, about 70km south-west of Hannover, combines comfortable accommodation in a modern building with fine dining in an historic restaurant dating from 1721. The business has been in the same

The DL2OBO Optibeam 17-4 Yagi.

family since 1756, more than 250 years, and is now run by Carsten-Thomas (Tom) Dauer, DL2OBO.

Tom has a fine shack that is available for use by radio amateurs staying at Hotel Hellers Krug. The station consists of a Yaesu FT-1000MP and FT-897, a Kenwood TS-940S, and Icom IC-7000 and IC-706 transceivers, to Ameritron AL-1200 and UY5ZZ 1kW amplifiers. Outside is an Optibeam 17-4 multiband Yagi with three elements on 40m, four elements on 20m, four elements on 15m and six elements on 10m. The tower is shunt-fed as a vertical on 80 and 160m. For receiving there is a two-wire Beverage system and a K9AY loop.

There is a lot to see and do in this part of Germany. The region is famous for its historic cities, ancient castles and places mentioned in well-known fairy tales such as the Pied Piper of Hamelin and Baron Muenchhausen.

The normal rates are EUR 82 euros

(approx GBP £59) for a double room, or 51 euros (approx £37) for a single room (including breakfast), but radio amateurs are given a special rate: please send Tom an e-mail to ask for details.
Contact: Carsten-Thomas (Tom) Dauer, DL2OBO, Hotel Hellers Krug, Altendorfer Str 19, D-37603 Holzminden, Germany.
Tel: +49 5531 2001;
fax: +49 5531 61266.
E-mail: carstendauer@hellerskrug.de

Ferienhaus Freimuth, Neufunnixsiel, East Friesland
Website:
www.ferienhaus-freimuth.de
Ferienhaus Freimuth is a guest house owned by VHF enthusiasts Reiner (DG7FX) and Sonja (DG6ZV) Freimuth,

Radio shack at Hotel Hellers Krug.

Antennas at Ferienhaus Freimuth.

at Neufunnixsiel near the North Sea coast of Germany. There are two apartments available for rent. One is for two to three people, the other (with two bedrooms) is for four or a maximum of five people.

There are horizontal Yagi antennas for 2m and 70cm as well as a duoband vertical for these bands. It is possible to work into Denmark, Sweden, the Netherlands, Belgium and England using 10W on 2m as the surrounding area is very flat. For the HF bands there is an FD4 Windom wire antenna and plenty of space for additional antennas should you wish to put up your own. Guests should bring their own transceiver.

The rates are from EUR 36 euros (approx £26) to 49 euros (approx £35) per apartment per night, depending on the length of stay and the season.
Contact: Reiner (DG7FX) and Sonja (DG6ZV) Freimuth, Kattrepeldiek 8, 26409 Wittmund-Neufunnixsiel, Germany.
Tel: +49 4464 942493.
E-mail: rsfreimuth@ewetel.net

GREECE

(SEE ALSO CRETE)
Haus Eisinga, Megali Mandinea near Kalamata, Peloponnese
Website: www.amateurfunkferien.de *and* **www.eisinga.de/aktuelles/aktuelles.htm (two different websites)**
Haus Eisinga is a large three-storey villa set 300 metres above the sea in the village of Megali Mandinea near Kalamata in southern Greece. It is owned by Jürgen (Joe) Eisinga, DL2YAG / SV0IE,

Haus Eisinga near Kalamata, Greece.

and his family.

Built in 1990 - 1991, the villa has a living area of around 300 sq m and includes an indoor swimming pool. The house has been divided into two apartments, and Apartment 1 includes the amateur radio station shack. Either Apartment 1, Apartment 2 or the whole house can be rented.

Apartment 1 is approximately 50sq m in size and includes a living room / kitchen with refrigerator, cooker and microwave oven. There is satellite TV with 106 channels, video player with 400 movies (in German), stereo with CD collection, library etc. The bathroom includes a washing machine. There is one bedroom with a king-size bed plus the shack, which can sleep two more people.

The shack is equipped with a Yaesu FT-901DM and Icom IC-726 HF transceivers, an Alpha 78 linear amplifier and a Yaesu FRG-7 receiver. For CW fans there is an ETM 4 keyer. For the

Hy-gain TH-7DX and spectacular view at Haus Eisinga.

VHF / UHF bands there are two Kenwood rigs: a new TM-742E three-bander and a TH-75E. The antennas include a Hy-gain TH-7DX 7-element beam for 10, 15 and 20m on a 10m tower on the roof of the villa, a G5RV, an Inverted-V, GPA-3 vertical, 2m Yagi, 2m ground plane, Diamond dual-band vertical and 70cm Yagi, and there are also plans for 6m and 23cm antennas. A Compaq Pentium II computer with high-speed Internet connection and wireless router can also be used at no additional charge.

Apartment 2 (with no radio shack) sleeps two people and also has a living room / kitchen and bathroom. The 8 x 4m indoor swimming pool is on the middle floor, with a 50 sqm terrace outside. A garaged Seat Cordoba may be used by guests by arrangement. At the back of the house is a terraced garden with olive grove (Kalamata olives and olive oil are among the best in the world).

Further information is on the additional website www.eisinga.de/aktuelles/aktuelles.htm which includes two short videos in .wmv format that can be downloaded (scroll right to the bottom of the page and click on 'Video - foto stories'). They are about 3.6 and 4.8MB in size and there are separate German and English language versions showing more of the village of Megali Mandinea and the villa.

Should further accommodation be required, there are two more apartments with one and two bedrooms at Haus Nicos, next door to Haus Eisinga. Pictures and a description in German are on the Internet at www.eisinga.de/aktuelles/aktuelles.htm - scroll about two-thirds of the way down the page and click on "Hier sehen Sie mehr über das Haus Nicos - See more about Nicos-House" on the left of the screen.

Contact: Jürgen Eisinga, DL2YAG, Königsberger Str 21, D-46238 Bottrop, Germany.
Tel: +49 2041 34106; **mob:** +49 172 2811262; **fax:** +49 2041 35657.
E-mail: juergen@eisinga.com
or
Jürgen Eisinga, SV0IE, Box 5, GR 24101 Megali Mandinea, Kalamata, Greece
Tel: +30 27210 58506,
mob: +30 69 4564 0254.

GUADELOUPE

DX Shack, Capesterre (IOTA NA-102)
Website: www.qth.com/dxshack/FG/FGtop.htm

The DX Shack on Guadeloupe was set up by Hiroo, XU7AAA (ex-JA2EZD), and is managed on the island by Georges, FG5BG. It is a simple one-bedroom building just 7m x 7m in size, set in a former banana plantation next to Roseau Beach with a great take-off across the Caribbean Sea. Capesterre Belle-Eau is the most developed tourist town in Guadeloupe.

The DX Shack is designed for a single operator or a maximum of two people sharing. If you are intending to operate with a larger group you should book an hotel for the other operators.

The DX Shack has two single beds in a small bedroom plus the radio room. Two fridges and a microwave oven are provided, with basic cutlery and crockery for two people. Outdoors on the terrace is a simple open kitchen with gas stove. There is a bathroom with shower and toilet.

The equipment provided includes a Kenwood TS-930 with CW filter, a Kenwood TS-450 also with CW filter, and an Icom IC-706 for HF plus 6m and 2m. There is an Alpha 76PA linear

A4S and Force 12 C4XL beams.

A beam's-eye view of the sea.

Contact: DX Shack FG Manager, Georges Santtalikan, FG5BG, Ilet Perou, F-97130 Capesterre Belle Eau, Guadeloupe, French West Indies. **Tel:** +590 86 4071; **mob:** +590 57 7438. **E-mail:** georges.fg5bg@wanadoo.fr

Beach near DX Shack with tower in background.

amplifier and a four-way antenna switch. You should take your own CW paddle, headphones and PC for logging. Antennas include a Force 12 C4XL beam for 10, 12, 15, 17, 20 and 40m on a 60ft telescopic aluminium crank-up tower as well as a Cushcraft A4S 4-element triband Yagi for 10, 15 and 20m on a 100ft Rohn tower. This tower also supports a 5-element Yagi for 6m and there are inverted-Vee dipoles for 160 and 80m.

The DX Shack is located in a tropical fruit garden and guests may help themselves to fresh fruit growing on the trees. To supplement your diet, you can buy imported American and European food at a supermarket 4km from the DX Shack or, if you prefer your meals to be provided for you, there is a restaurant just 70m away, and another two restaurants 2.5km from the shack.

If you stay more than a day or two you are advised to hire a car, as the location of the DX Shack is fairly remote. There are many rent-a-car offices at Point-a-Pitre airport, 50km away, and rental is about USD $300 per week for a 1200cc car.

Further details can be obtained from the DX Shack FG Manager, Georges Santtalikan, FG5BG. You can write to Georges in French, Créole, English or Spanish.

GUINEA (REPUBLIC)

and

GUINEA-BISSAU

4WD Portable DXpedition

Jan-François, 6W7RV, offers a portable DXpedition in his 4X4 vehicle equipped with radio from 80m to 70cm. The vehicle has two 105Ah batteries with 400/800W 12V - 220V inverter and is equipped with an SGC SG-230 SmartTuner for 1.8 - 30MHz (max 200W) with SG-202 antenna, and Maldol HFC-40L, HFC-30L and HFC-20L mobile antennas for 40, 30 and 20m. The 4X4 vehicle has a special 50 litre water tank and a 45 litre fridge. Guinea and Guinea-Bissau are two of several possible destinations. For further details see the main entry under SENEGAL.

HAWAII

**Leilani bed & breakfast, Big Island, Hawaii (IOTA OC-019)
Website:
www.leilanibedandbreakfast.com**
Off the beaten track, Leilani Bed & Breakfast caters to the adventure traveller visiting the Big Island of Hawaii

The shack at Leilani B&B, Big Island, Hawaii.

perfect year-round climate. This radio shack has excellent signal reception with a low-noise level. You won't be disappointed with the big signal from the Big Island.

Personal coaching, massages and an exercise bike are available and there is even an art gallery on the premises.

Leilani Bed and Breakfast is close to three areas to view whales from the shoreline or, for even closer encounters, local boat excursions are available. Deep-sea (sport and marlin) fishing is another possibility. Bird watching is a treat and a wide variety of species visits Leilani every day. The location is convenient for visiting the Hawaii Volcanoes National Park, the number one tourist attraction in Hawaii.

Three 'themed' rooms are available, each with private facilities: the Orchid Room and the Shell Room each sleep two and cost from USD $75 plus local tax per night, while the Palm Room can sleep up to four in a queen bed and two twin beds and costs from $85 plus lo-

who wants comfortable, relaxing and affordable accommodation. The house is built of rock and has a rustic look. It is set amongst tropical gardens and fantastic volcanic formations: the property borders three huge lava tubes (pukas). A 900 sq ft enclosed garden lanai with waterfall is a special feature.

Macadamia nuts, figs, papayas, pineapples, asparagus, coffee, oranges, lemons, strawberry guava and kumquats are grown on the property and Leilani Bed & Breakfast uses its own fruits and nuts in season for breakfast.

There is a separate ham radio building, fully equipped with state of the art amateur radio equipment and antennas. This includes an Icom IC-756ProIII, an Ameritron AL-82 legal-limit amplifier, and a SteppIR 4-element Yagi on an aluminium tower. A full complement of Heil headsets and microphones plus other essential gear are available for your use. You may also use your laptop on our wireless Internet connection. You may e-mail for a schedule to hear this station on the air.

Operate this station comfortably in undisturbed total peace and quiet from this island location with cal tax per night. Rental of the radio shack starts at $150 per 24-hour operation. Special weekly package prices including B&B rooms are available: please enquire by e-mail. Prices are subject to change.

The property is listed in the *Lonely Planet* travel guide.

Contact: Randy (KH6RC) and Lynn Van Leeuwen, Leilani Bed & Breakfast, PO Box 7209, 92-8822 Leilani Parkway, Ocean View, HI 96737, USA.
Tel: +1 808 929 7101;
fax: +1 888 486 6660.
E-mail: leilanibnb@fastnethi.com

KD4ML Ham Radio Studio, North Kohala, Big Island (IOTA OC-019)
Website: www.qrz.com/kd4ml

A clean, comfortable ham radio studio located on the northern tip of the Big Island, the KD4ML station is equipped with a Kenwood TS-50, matching auto tuner and a choice of antennas. There is a wi-fi Internet connection available.

Accommodation is in a 400 sq ft guest suite with queen size bed, private bathroom, sitting area and tropical deck entrance. Cable TV, fridge, microwave oven, coffee maker and toaster are included.

The weather in North Kohala is cooler and breezier than the Kona Coast and the trade winds usually keep temperatures comfortable. Non-smokers only and, unfortunately, accommodation is not available for children.

Rates are USD $89 per night for a one or two-night stay, $79 per night for minimum three nights, or $499 for a week.

Contact: Howard (KD4ML) and Carol Ann Olsen, PO Box 795, Kapaau, HI 96755, USA.
Tel: +1 808 889 5431.
E-mail: kd4ml@juno.com

Sea Q Maui, Pukalani, Maui (IOTA OC-019)
Website: www.seaqmaui.com*

Sea Q Maui is an amateur radio rental station that is only open to radio amateurs and their families. Anyone with an amateur radio licence may stay there and almost any licensed ham from any country may operate from the *Sea Q Maui* station. The fees charged are to pay for the upkeep of the station and antennas and of course the use of the accommodation.

Many contests have been won from the *Sea Q Maui* station and many who have operated from here return. Antennas include (on tower one) a Cushcraft A4S 4-element beam for 10, 15 and 20m and an A3WS 3-element Yagi for 12 and 17m, and (on tower two) a Cushcraft 2-element 40m beam and a Hy-Gain TH-7DX 7-element Yagi for 10, 15 and 20m.

Sea Q Maui is a large Hawaiian-style home featuring vaulted ceilings, spacious *lanais*, and ocean views. It is situated adjacent to the 15th fairway of the Pukalani golf course and has unrestricted views and beautiful gardens. Located at 1200 ft ASL it is cooler and much less congested than in town, yet close to everything.

The area is located near the Haleakala Volcano National Park, tropical rainforest, Hookipa Beach (known as the windsurfing capital of the world) and excellent dining.

Guests may do their own cooking but the host likes you to have breakfast with him (included).

Three rooms are available, all with private bathroom facilities, but generally there are only two people staying as guests as there is only the one station and *Sea Q Maui* is not open to non-hams.

Contact: Terry Clayton, KH6SQ, 255

Kenwood TS-50 at KD4ML studio.

Antennas at Sea Q Maui.

Kaualani Dr, Pukalani, HI 96768, USA.
E-mail: kh6sq@arrl.net
* Please note that this website is password protected. Go to www.seaqmaui.com for instructions on how to get in. It is open only to hams with a valid callsign.

ITU GENEVA

4U1ITU club station, ITU headquarters, Geneva
Website: http://life.itu.int/radioclub/index.htm
Anyone who has a current amateur radio licence may operate the 4U1ITU station at the headquarters building of the International Telecommunication Union. Operation is possible on all Region 1 frequencies between 1810kHz and 440MHz. Authorisation must be obtained well in advance of your visit. Without such authorisation it is not possible to enter the ITU HQ building. **Contact:** International Amateur Radio Club, PO Box 6, CH-1211 Geneva 20, Switzerland.
E-mail: 4u1itu@itu.int

JAMAICA

6Y1V contest station, Hopewell, near Montego Bay (IOTA NA-097)
The 6Y1V contest 'superstation' is located in the hills just above the coastal village of Hopewell on Jamaica's northwest coast, 17km west of Montego Bay. The location is about a mile from the sea at 850ft above sea level and has a panoramic view of the Caribbean Sea from west through north to east.

The 6Y1V station is designed to operate a variety of amateur radio contests in any category, including single operator, SO2R, multi-operator, multi-operator single transmitter, multi-operator two transmitter and multi-operator multi-transmitter. The station is also designed to be competitive in all power categories including QRP. The radios are four Icom IC-7800 transceivers, one Yaesu FT-1000MPMkV (fully loaded with filters and mods), and a Kenwood TS-2000 used primarily for VHF. Each of the IC-7800s can be used independently or

135ft Rohn tower with MonstIR Yagis at 70ft and 130ft.

paired for SO2R operation. Each pair of IC-7800s shares an Acom 2000A linear amplifier. The Yaesu can use either a Kenwood TL-922 or an Ameritron AL-80B (the TL-922 and AL-80B will soon be replaced with another Acom 2000A and an Acom 1000 for use on the 6m band).

There are currently six towers and plans for a few more. The first tower is a 135ft Rohn 45G supporting two Fluid Motion SteppIR MonstIR Yagis at 70ft and 130ft. The MonstIR Yagis are stacked and phased and are designed mainly for run station use. They can provide 4-over-4 elements on 10, 12, 15, 17 or 20m, or 3-over-3 elements on 30 or 40m. They can also be rotated and used independently for multi-multi operation. These antennas are great for WARC band operation and are also configured to operate as 6-element Yagis on 6m.

Tower number two is a 105ft Rohn 45G rotating tower. The rotating tower supports six M² long-boom monoband Yagis. On 20m there is a 6-over-6 array on 59ft booms at 50 and 100ft. On 15m, another 6-over-6 array at 32 and 65ft, and 7-over-7 array for 10m at 28 and 55ft. The primary purpose for the rotating tower and stacked arrays is for multiplier hunting: the antennas have a razor sharp beamwidth and provide maximum gain.

There are four Hy-Gain Hy-Towers extended to 67ft for 80m. These provide an excellent four-square array with >25dB of front-to-back ratio. On 160m there are four slopers from the top of the 135ft tower. For low-band receiving, there are two 1000ft Beverages, one NW and one NE, a K9AY loop and a Pennant.

The computers are in a Windows domain and connected via a Cisco network with a PIX firewall. Guests simply log in and select the logging program of choice and operate. For convenience, log files are automatically backed up over remote VPN to a data centre in Louisville, Kentucky. Internet browsing, telnet and checking e-mail is all permitted.

The entire station can be remotely controlled over the Internet using VPN technology and VoIP. Each piece of equipment can be remotely powered and is backed up by UPS. There is an on-site generator for emergency power.

You may rent the station (with 24-hour access) and stay at any of the local resorts, or rent the station and house together. All rentals must be booked through David, KY1V. Rental for

One of the extended Hy-Gain Hy-Towers used in a 4-Square array on 80m.

6Y1V antenna panorama.

the station only is from USD $250 per day or $1500 for a non-contest week, up to $3500 including rental of the house for a week that includes a major contest. A $1500 refundable deposit is required for weekly rentals and a $500 deposit required for daily use. If you wish to visit the station for a short tour, please make a reservation in advance. Arrangements can be made for you to operate for a few hours at no charge. (Please note that station owners KY1V and K1LZ receive no funds from rental of the station. Rental proceeds are used entirely for station improvements, equipment maintenance, maintenance and improvements at the house, paying for the Internet service and a small salary for the station custodian, George Campbell, 6Y5GC. George may also act as a guide for visitors upon request; tipping is customary).

If travelling to Jamaica to visit 6Y1V, you should fly to Montego Bay and not Kingston, which is more than a five-hour drive away. During heavy rains, you may not be able to cross the mountains.

Contact: David Kopacz, KY1V, 2920 Cox Mill Road, Hopkinsville, KY 42240, USA.

Tel: +1 615 345 6660.

E-mail: david@ky1v.com

Jamaican Retreat, Malvern (IOTA NA-097)
Website: www.infochan.com/~joshwa

Approximately three miles inland from the south coast near Malvern, Jamaican Retreat is located on the south face of the Santa Cruz mountain range. At 2300ft ASL the air is cool and there are panoramic views of the lowland areas and the ocean.

Self-catering accommodation is offered in a detached chalet with *en suite* facilities, double bedroom, small bunk bedroom, lounge and kitchen. There is a large roof-top type sun lounge from where to relax and take in the view, or a patio on which you can sample the delights of Jamaican home cooking.

Bedroom at Jamaican Retreat visitor's chalet.

Jamaican Retreat radio shack.

You are welcome to take along your own equipment. The electricity supply is both 240 and 110V at 60Hz, and there is a 3.5kW generator for back-up. The antenna system is a 2-element cubical quad for 10, 12, 15, 17 and 20m. Owner of the Jamaican Retreat, Josh Walker, 6Y5WJ, said, *"I'm no longer offering the use of the radio equipment, and I'm in the process of renovating the antenna system, i.e. 5-band quad on an automatic lift tower. In the past ham visitors have preferred to bring their own equipment - less DX time wasted using an unfamiliar rig, especially during competitions. If it's just casual DX, then on arrival I'll set up one of the rigs."*

Telephone and Internet facilities are available and there are cellular phone systems in operation on the island.

The surrounding farmland provides fresh vegetables of all types and there is an abundance of fruit to sample and enjoy. Josh and his wife can provide traditional Jamaican dishes or there are local licensed restaurants offering a wider variety of menus along with wines and Caribbean beverages or punches.

The rates range from £250 per week during the low season to £350 a week in the high season for self-catering, and they include all electricity and gas usage. Alternatively, all meals and housekeeper services can be provided at a flat rate of £10 per day. (Note: all prices are correct as of 15 December 2007.)

Contact: Josh Walker, 6Y5WJ, Box 109, Southfield PO, St Elizabeth, Jamaica. **Tel:** +1876 448 2932 *or* +1876 360 2819 from 0900 - 2100 local time (1400 - 0200UTC).

E-mail: joshwa@cwjamaica.com

JERSEY

Jersey Amateur Radio Society club station, GJ3DVC, La Moye (IOTA EU-013)
Website: www.gj3dvc.org.je

The Jersey Amateur Radio Society (JARS) club station is located in a WWII German signal station at La Moye in the parish of St Brelade, about 3 miles west of the capital of Jersey, St. Helier.

The club shack is equipped with a Yaesu FT-1000 transceiver, a 3-500Z linear amplifier for 10 - 80m, a 3-element triband beam, a 40 / 80m dipole, a 160m dipole, a Cushcraft R8000 vertical and 2m FM capabilities. (The equipment available may vary, but this is indicative of what is available.) There is an area of land that can be used for wire antennas or temporary masts and also a large flat roof area.

The JARS shack may be rented, subject to the following terms and conditions. The shack is available on a first-come first-served basis. Confirmed reservations will be filled first. Use of the club station is only available to club members. Overseas membership is available to non-residents of Jersey, currently at £5.00 per year. A rental charge is payable for each operation by overseas members. The electricity meter will be read upon arrival and prior to departure and the electricity used must be paid for.

A booked operation may ordinarily consist of up to seven consecutive days.

Jersey Amateur Radio Society Club Station.

From persons not known to the club a deposit of £200 is required. This will be returned following an inspection to ensure the club is left in a clean and tidy condition and all equipment is connected and operational as it was prior to the operation. Any damage must be repaired or paid for.

It is strictly forbidden for any visitors to tamper with the radio or aerial installations in any way. An ADSL connection is available in the shack. It is not permitted to alter any of the settings in the router for any reason.

Accommodation in an hotel, guest house or camp site must be booked for the duration of your stay. The club house is not permitted to provide accommodation for visitors.

Unless an arrangement has been made in advance, priority is to be given to club members who may wish to use the shack on club nights (currently Wednesday and Friday 8.00pm to 10.30pm).

If any of the club callsigns (GJ3DVC, GJ2A or GJ8RVT) are used, an electronic copy of the log must be e-mailed to the club. The GJ2A callsign may only be used by club members and is only available for use during certain specified contests. It may not be used before or after the contest.

Contact: The Secretary or President, The Jersey Amateur Radio Society, The Old German Signal Station, Le Chemin des Signaux, St Brelade, Jersey, Channel Islands, Great Britain.
E-mail: gj3dvc@gj3dvc.org.je

LESOTHO

African DX Safaris

African DX Safaris can organise tailor-made DXpeditions to Lesotho, among other countries, providing accommodation, meals, equipment and antennas. See main listing under SOUTH AFRICA.

MACEDONIA

Z36A station, Kocani
Website: www.qrz.com/z36a

Zoran Levkov, Z36A, offers the use of his station in Kocani, in the eastern part of Macedonia. The station is in an

Zoran, Z36A, at his station.

apartment on the top floor of a 20m-high building. On the roof is a Hy-Gain TH3Mk3 beam for 10, 15 and 20m on a 6m tower. A Fritzel rotary dipole for 12, 17 and 30m is mounted under the beam and there are also dipoles for 40 and 80m. A 6m 3-element ZX-Yagi is mounted on the highest part on the building and there is a 2m 5/8-wave vertical for local chats.

Inside the shack is a Kenwood TS-930S and an Icom IC-706MkII, with a Heathkit SB-200 linear amplifier. Internet access is also available.

The apartment can accommodate up to six people. Further details can be obtained from Aco Jevremov, DJ0LZ, in Germany.

Contact: Aco Jevremov, DJ0LZ.
E-mail: ace.jevremov@googlemail.com

The 10m 6-element Yagi at Z37M (see page 119).

The 'Nikola Tesla' Radio Club, Z37M, in Stip, Macedonia.

Z37M contest station, 'Nikola Tesla' Radio Club, Stip
Website: www.qsl.net/z30m

Z37M is one of the best-known contest stations in Macedonia. It is located at the 'Nikola Tesla' Radio Club in Stip in the eastern part of Macedonia.

Accommodation is available for up to six people.

Equipment in the shack includes a Yaesu FT-1000MP, a Yaesu FT-920, a Kenwood TS-940SAT, a Kenwood TS-850SAT, a Kenwood TS-430S, and an Icom IC-745 to a Dentron MLA-2500 and three home-made 700W linear amplifiers. A 24-hour ADSL Internet connection is available.

The antennas include an inverted-V for 160m, a half-square for 80m, two delta loops for 80m, a half square for 40m, a delta loop for 40m, a 4-element Yagi for 20m on a 15m high tower, a 5-element Yagi for 15m on a 12m high tower, a 6-element Yagi for 10m on a 12m high tower, a Hy-Gain TH6DXX triband beam, a 2-element 5-band cubical quad (20 - 10m) and a vertical for 30m. There are also Beverages and K9AY receiving antennas.

Enquiries about visiting the Z37M contest station should be made to the Vice-President of the 'Nikola Tesla' Radio Club, Venco Stojcev, Z36W.
Contact: Venco Stojcev, Z36W.
E-mail: z36w@mt.net.mk *or* z36w@t-home.mk

MADAGASCAR

La Villa Suède, Antananarivo (IOTA AF-013)
Website: http://pejy.simicro.mg/villarose

Comfortable rooms in a quiet area of Antananarivo are offered by Åke, 5R8FU, and his Madagascan wife, Jacqueline, at their guest house 'La Villa Suède'. There are four rooms and an apartment available for rent. Breakfast is included in the rental and lunch and dinner are also available if required.

The station includes a Yaesu FT-

80m vertical at Z37M contest station.

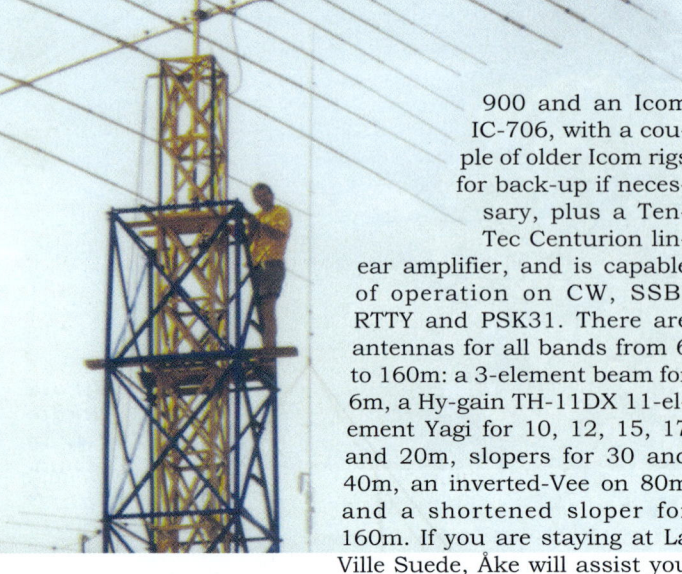

Åke, 5R8FU, working on the antennas.

900 and an Icom IC-706, with a couple of older Icom rigs for back-up if necessary, plus a Ten-Tec Centurion linear amplifier, and is capable of operation on CW, SSB, RTTY and PSK31. There are antennas for all bands from 6 to 160m: a 3-element beam for 6m, a Hy-gain TH-11DX 11-element Yagi for 10, 12, 15, 17 and 20m, slopers for 30 and 40m, an inverted-Vee on 80m and a shortened sloper for 160m. If you are staying at La Ville Suede, Åke will assist you in obtaining a Madagascan licence and an import permit if you wish to bring in your own equipment.

Åke says, *"Most of my ham visitors have not done much else but working pile-ups. But if you take the trouble to come all the way down here, you should take the time to do other things as well. And there is certainly much to do here in Madagascar. It is a very large island with unique flora and fauna. There is a number of wildlife parks where you can stay and wander around. Or you can go down rivers and canals. There are many alternatives. Of course, there are also thousands of miles of nice beaches for swimming, diving and relaxing."*

At present Åke and Jacqueline can take visitors on day trips around the Antananarivo area, but they hope to be able to expand this side of the business soon in order to take visitors on longer tours themselves. Meanwhile, they can arrange sightseeing tours to other parts of the country.

The rates are EUR 25 euros (approx GBP £18) per room per day, or 40 euros (approx £28.50) per day for the apartment, including breakfast for two people. Airport transfers can be arranged for a very reasonable sum.

Contact: Åke, 5R8FU, and Jacqueline Rosvall.

Tel: +261 32 04 656 66
or +261 20 22 523 97.

E-mail : villarose@netclub.mg

MALAWI

Red Zebra Lodge, Kambiri Point, Senga Bay, nr Salima
Website: www.lakemalawi.com (see also http://malawi.g4axx.com)

Red Zebra Lodge is at Kambiri Point right on the lakeshore in Senga Bay, 20km from Salima town and 110km from Lilongwe international airport (one and a half hours on an excellent newly-paved road).

Tropical fresh-water cichlids are caught in Lake Malawi for export to aquariums and collectors all over the world and the lodge is located at the site of Malawi's biggest ornamental fish exporting business. It gets its unusual name from a species of cichlid.

Red Zebra Lodge consists of eight double rooms each *en suite* with shower, toilet and wash basin with hot and cold running water. The rooms have one double or two single beds with mosquito nets and a oscillating fan. Six rooms are in a modern accommodation block while the remaining two are in an older African-style thatched guest cottage with views over the lake.

Meals are served in a thatched

La Villa Suède, on the outskirts of Antananarivo.

PHOTO: 9M6DXX

Assembling the Cushcraft A3S.

rondavel and are prepared by experienced cooks. Every opportunity is taken to provide visitors with Lake Malawi fish dishes - usually *chambo* (tilapia) or *kampango* (lake catfish). Soft drinks, beer, spirits and wine are available from the bar in the *rondavel*.

There is no permanent amateur radio station set up at Red Zebra Lodge, but the location has been used by a small number of DXpeditions and the lodge staff are keen to make the site available to more amateur radio operators in future. The DXpeditions have left behind what now amounts to a considerable amount of equipment, e.g. the 2004 7Q7MM DXpedition left a new Cushcraft A3S 3-element Yagi for 10, 15 and 20m as well as coaxial cable and polypropylene guy ropes. The A3S is neatly secured in a ski bag with its instructions, and lengths of RG213 and RG58 coax are carefully stored in sealed drums. You will need to assemble and put up the beam yourself, as well as take it down, dissemble it and put it back into storage at the end of your operation, but you will find plenty of enthusiastic helpers at the lodge ready to give a hand!

A complete inventory of equipment either left behind by DXpeditioners or owned by the lodge and made available to radio amateur groups was compiled by the 7Q7MM team and this may be found on their website at http://malawi.g4axx.com/Inventory.php While no guarantees can be given as to its present availability or condition all the equipment was in good working order when packed away in 2004. You should take your own transceiver, lin-

PHOTO: 9M6DXX

ear amplifier (if required), coax patch leads, PC for logging, keyer, headset etc.

Many sightseeing excursions can be arranged, from an interesting walk-about tour of the ornamental fish exporting facility adjacent to the lodge to a four-day three-night visit to the South Luangwa National Park across the border in Zambia. Divers can explore Lake Malawi and observe the underwater habitat of the cichlids. A day trip on the Shire river into the Liwonde National Park offers the chance to see elephants, hippo, crocodiles and many species of birds and other game. See the www.lakemalawi.com website for further sightseeing suggestions.

Current rates (as of June 2008) are: bed and breakfast (B&B) (single) USD $33.00 per night; B&B (double) $60.00; B&B + dinner (single) $50.00; B&B + dinner (double) $85.00; full board (single) $60.00; full board (double) $105.00. **Contact:** Esther Grant, Red Zebra Lodge, PO Box 123, Salima, Malawi, Central Africa. **Tel:** +265 1 263 165; **fax / tel:** +265 1 263 407; **mob:** +265 9 233 428 *or* +265 9 913 630. **E-mail:** esther@lakemalawi.com

Operating shack and Cushcraft A3S beam at Red Zebra Lodge, Malawi.

PHOTO: 9M6DXX

Esther Grant with her late husband, Stuart.

MALI

4WD Portable DXpedition

Jan-François, 6W7RV, offers a portable DXpedition in his 4X4 vehicle equipped with radio from 80m to 70cm. The vehicle has two 105Ah batteries with 400/

800W 12V - 220V inverter and is equipped with an SGC SG-230 SmartTuner for 1.8 - 30MHz (max 200W) with SG-202 antenna, and Maldol HFC-40L, HFC-30L and HFC-20L mobile antennas for 40, 30 and 20m. The 4X4 vehicle has a special 50 litre water tank and a 45 litre fridge. Mali is one of several possible destinations. For further details see the main entry under SENEGAL.

MARTINIQUE

The Madinina Shack (IOTA NA-107)
Website:
http://pagesperso-orange.fr/fm5bh

The Madinina Shack, location of the FM5BH contest station on the island of Martinique, is available for rental by radio amateurs. The station includes a Yaesu FT-1000MP with CW filter, a Kenwood TS-930S and an Alpha 91b linear amplifier to a Force 12 C31XR beam and a Cushcraft 40-2CD 2-element Yagi for 40m on a 22m tower. A Cushcraft A4S 4-element Yagi for 10, 15 and 20m at 14m is also available. For the low bands, there is an inverted-V dipole for 80m and an inverted-L for 160m. Also in the shack is a computer loaded with various logging programs etc.

The rental shack is housed in a one-bedroom chalet with air conditioning. It includes a bathroom with hot shower and a kitchen with refrigerator and microwave oven.

The station is located 15 minutes

Madinina station, Martinique.

from Martinique's beaches and a short walk from shops and restaurants. The rate is EUR 80 euros (approx GBP £58) for two people per night, with a minimum rental period of five nights.
Contact: Laurent Bellay, FM5BH, 1 Allée des tourterelles, 97224 Ducos, Martinique.
Tel: +596 561990.
E-mail: fm5bh@wanadoo.fr

MAURITANIA

4WD Portable DXpedition

Jan-François, 6W7RV, offers a portable DXpedition in his 4X4 vehicle equipped with radio from 80m to 70cm. The vehicle has two 105Ah batteries with 400/800W 12V - 220V inverter and is

The 6W7RV radio-equipped 4WD at a camp site in Mauritania.

Laurent, FM5BH.

equipped with an SGC SG-230 SmartTuner for 1.8 - 30MHz (max 200W) with SG-202 antenna, and Maldol HFC-40L, HFC-30L and HFC-20L mobile antennas for 40, 30 and 20m. The 4X4 vehicle has a special 50 litre water tank and a 45 litre fridge. Mauritania is one of several possible destinations. For further details see the main entry under SENEGAL.

MAURITIUS

3B8CF beach apartment, Flic en Flac (IOTA AF-049)
Website: www.qsl.net/3b8cf

Jacky, 3B8CF, has two rental stations available for use on Mauritius. The first is a self-catering apartment just 50m from the beach at Flic en Flac on the west coast of Mauritius. Here, there are three rooms with all amenities. The apartment is suitable for up to six people, although the best arrangement is for up to four people, leaving the other room free for use as the shack. There is air conditioning in two of the rooms. The radio room has a 12V PSU provided and the antenna is a Cushcraft R7 vertical on the roof. It is also possible to put up some wire antennas.

The rental charges for the beach apartment are also negotiable and arranged when booking.

Contact: Seewoosankar 'Jacky' Mandary, 3B8CF, Post Office Box 104, Quatre Bornes, Mauritius.
Tel: +230 424 5866;
mob: +230 763 6713.
E-mail: 3b8cf@intnet.mu

3B8CF Beach Apartment, Mauritius.

3B8CF station, Quatre Bornes (IOTA AF-049)
Website: www.qsl.net/3b8cf

Jacky's second station is an apartment with radio shack at his home on an elevated site in the town of Quatre Bornes. The apartment has a bedroom with bedding provided, kitchenette with fridge, toilet, shower room with towels provided, and a radio room with 12V PSU.

The antennas are a Cushcraft A4S four-element Yagi for 10, 15 and 20m, a Cushcraft A3WS three-element Yagi for 12 and 17m, a 30 / 40m rotary dipole and inverted-Vees for 80 and 160m. At certain times power has to be reduced to around 30W, especially on 17, 15 and 12m, due to TVI problems.

Mauritius is small and nowhere is that far away, so all tourist places are easily accessible. The nearest beach is 18km away at Flic en Flac on the west coast of the island. The price is negotiable, and agreed upon reservation.
Contact: Seewoosankar 'Jacky' Mandary, 3B8CF, Post Office Box 104, Quatre Bornes, Mauritius.
Tel: +230 424 5866;
mob: +230 763 6713.
E-mail: 3b8cf@intnet.mu

The 3B8CF Station at Quatre Bornes, Mauritius.

Living room at the 3B8CF guest apartment.

403A: one of the best sites in Europe.

MONTENEGRO

Premium Contest Resort, 403A station, Herceg Novi, Montenegro Website: www.yt6a.com

Yes, a great DX shack location is available for rent in one of the world's newest countries! The station has been built up by owner Ranko Boca, 403A (formerly YT6A), with top-level HF contesting in mind. Now, a three-bedroom, two-bathroom apartment with living room and kitchen is available for rent at the hilltop site. Electricity comes from a 25kW generator. Meals are prepared for paying guests by an employee on site.

Sunset at the 403A contest 'superstation'.

The station is suitable for single or multi-operator DXpedition-type activity or contests (SO2R or multi-single are

Antenna work at 403A.

immediately available, for multi-multi discuss with Ranko). The antenna farm is constantly being upgraded. In February 2008 there were four towers supporting the following antennas: 160m dipole, 80m 2-element Yagi, 40m 3-element Yagi, 20m 3-over-3-over-3 and 3-over-3 stacks and separate 3-element Yagi, 15m 4-over-4-over-4 and 5-over-5 stacks, 10m 5-over-5 stack, plus three Beverage antennas for low-band reception.

Two airports are very close. Dubrovnik in Croatia is 25km away and has very good connections, while Tivat airport in Montenegro itself is just 10km away.

The 403A 'shack'.

Ranko Boca, 4O3A, on the air.

The daily price, with meals, is EUR 85 euros (approx GBP £61) per person. For major contest weekends the cost is 200 euros (approx £144) per person for Friday, Saturday and Sunday. Contact Ranko Boca by e-mail for further details. **Contact:** Ranko Boca, 4O3A, Nikole Ljubibratica 78, Herceg Novi, Montenegro.

E-mail: yt6a@cg.yu

MONTSERRAT

VP2MDD Villa at Old Towne (IOTA NA-103)

Graham Dawes, VP2MDD / M0AEP, owns a bungalow-style house and apartment in Old Towne, located on a hillside on the west coast of Montserrat, overlooking the Caribbean Sea and about 30 minutes drive from the new airport.

The house, which is built out of a steep hillside, offers a large dining/sitting area, two *en suite* bedrooms, a small office, a kitchen fitted with microwave oven, electric kettle, hob, fridge freezer, toaster etc, and a utility room. A self-contained twin-bedded apartment is under the main house and there is an outdoor swimming pool.

Telephone and broadband Internet are provided

and cable TV can be hired. The house has 240V AC with UK-style wall sockets and one 110V USA-style outlet in the utility room. Outside, at the end of the house, is an 18ft lattice mast with an extendable steel pole. Antennas (with coax provided) are an Eagle 5-element 6m Yagi and a 6m Cushcraft Ringo vertical, with a Carolina Windom for 10 - 80m. A degree of fitness is required to mount the Yagi as the mast has to be climbed and a short step taken on to the shallow roof to put the Yagi and vertical on the pole. The antenna is rotated manually. The proximity of a mahogany tree restricts the boom length of the antenna to about 20ft.

Montserrat's volcano, Souffriere, is four miles south of the house but with the Belham Valley between the two, the Montserrat Volcano Observatory assesses the risk as low. Although quiet

Mast with Carolina Windom and 6m beam.

Villa at Old Towne, Montserrat.

Another view of the wire antenna and 6m beam.

at the time of writing, Souffriere can erupt explosive ash and pyroclastic flows with little or no warning, so a portable radio capable of receiving news updates from local radio station ZJB on VHF / FM is essential. Graham says, *"The Belham River valley has been our saviour so far and pyroclastic flows have not come down or crossed it yet. All part of life in Montserrat, although thankfully everything has been very quiet for last 12 months."* Graham also says he would be pleased to help those renting his house to obtain a VP2M licence.

Hire car is desirable although taxis aren't expensive, e.g. about £10 for the 30-minute journey from the airport to the house. Montserrat is 20 minutes flying time, in a 19-seat aircraft, from the international hub in neighbouring Antigua.

The house is rented for GBP £500 per week in the low season (16 April to 14 December) and £600 per week in the high season (15 December to 15 April). A maid service is available if required at additional cost.
Contact: Graham Dawes, M0AEP / VP2MDD, Middlegate House, The Wold, Worlaby, Brigg, North Lincolnshire DN20 0NP, England.
Tel: +44 7790 618658.
E-mail: vp2mdd@gmail.com

MOROCCO

CN2DX beach house, Dar Bouazza, near Casablanca
Website: www.maroc-beach-house. com/index-ham.html
André Breguet, HB9HLM / CN2DX, offers his apartment at Dar Bouazza, 20km south of Casablanca and five minutes walk from the Atlantic Ocean.

Apartment F3 in the Beach House complex has two bedrooms, living room, kitchen and bathroom. There is a swimming pool, gym and restaurant in the Beach House complex.

You should bring your own transceiver, headphones, keyer etc, but a 25A 13.8V PSU, ATU (maximum 250W), HF and VHF SWR / power meters and various cables and mains adaptors are provided. Proximity to the sea has caused problems with oxidization of the HF and VHF beams, but as of 2008 a G5RV for 10 - 80m, a 160m dipole and a vertical for 6m, 2m and 70cm are available for use.

The rental is EUR 70 euros (approx £50) per day and the apartment is easily reached from Casablanca's Mohammed V international airport.
Contact: André Breguet, HB9HLM, Gare 49, 2017 Boudry, Switzerland.
E-mail: hb9hlm@net2000.ch

MOZAMBIQUE

African DX Safaris
African DX Safaris can organise tailor-made DXpeditions to Mozambique, among other countries, providing ac-

PHOTO: 9M6XRO

DX Safaris' Trailer-mounted Tennadyne T-8 log-periodic Yagi at Blue Anchor Inn, Mozambique.

PHOTO: 9M6XRO

DX Safaris' Cushcraft A4S and A3WS beams at Blue Anchor Inn.

commodation, meals, equipment and antennas. See main listing under SOUTH AFRICA.

NETHERLANDS ANTILLES

NETHERLANDS ANTILLES - BONAIRE (PJ4)
Noah Gottfried, K2NG, house (IOTA SA-006)
Websites:
http://dxholiday.com/sa/pj2.htm,
www.dxholiday.com/sa/k2ng.htm
Noah Gottfried, K2NG, offers a new rental station on the island of Bonaire.

Working on the F12 240N and C3.

The house was built in 2000 and is around 1000 square feet in size. It stands on a quarter acre plot with a 60ft tower and behind the house is an additional one acre plot with a 90ft tower. There are three air conditioned bedrooms with a queen sized bed in each. The master bedroom has an attached bathroom and there is a second bathroom shared with the other rooms. There is an open living space with a cathedral ceiling and two ceiling fans. The open living space has the shack table, dining area, kitchen and a futon that can be used as a couch or fourth bed. There is also a utility room with storage and a washing machine.

An Icom IC-756Pro and two Ameritron AL1200 linear amplifiers are provided for the use of the renters, plus an old Dentron MLA-2500. Renters should take all their own interconnect cables, keys, keyers, SWR meter, tools, computers, etc. It is also recommended that renters should take their own

Looking down from a 140ft cell tower to 60ft tower with C31XR.

K2NG working on the Force 12 C31XR at 60ft.

rig because it cannot be absolutely guaranteed that the IC-756Pro will be working.

The Yagis are by Force 12: a C31XR on the 60ft tower, a 240N (2-elements on 40m with 56ft elements) with a C3 tribander mounted above on the 90ft tower. An 80m dipole hangs between the towers at an average height of about 75ft. A 160m ground plane is mounted on the side of the 90ft tower. There is a pulley at the top of the 90ft tower for putting up temporary wire antennas as well as feedlines for Beverages or other antennas. In the storage room there is a Cushcraft A3 tribander and a 6m 3-element beam that can be put up temporarily. A climbing belt is available to renters.

There is also a 5kW generator available for back-up power to the house and station. This is a good feature for serious contest groups who want to make sure they complete a contest even if the power goes out. The power in Bonaire is generally reliable but there has been the occasional major power interruption.

The rent for 2008 is USD $1500 per week with one week minimum rental for all major contests. There is also a per person per night room tax (which goes to the Bonaire government).

View from cell tower looking down on the 90ft tower with F12 240N (2-ele on 40 with 56ft elements) and a C3.

The renter must also pay a $100 per week equipment usage fee to the PJ4 Radio Club to help pay for the equipment upkeep. All major contests require a deposit to reserve. If the renter must cancel his deposit will be refunded if a replacement renter is found. For small contests or non-contest rentals, it may be possible to rent the house for periods of less than a week on a per case basis. The rent in 2009 and beyond will probably go up to $1750 a week.

To check availability, contact owner Noah Gottfried, K2NG, but to make reservations or if you have questions about visiting Bonaire, contact Bert at Bonaire partners. Bert and his Bonaire Partners staff look after the rental of the house, can arrange rental car reservations and help out with any other vacation needs.

Contact: Noah Gottfried, K2NG, PO Box 481, Johnsonburg, NJ 07846, USA.
Tel: +1 845 731 2203 (7.00am - 3.00pm EST, 1200 - 2000UTC);
Mob: +1 908 268 5162.
E-mail: noah.gottfried@us.fujitsu.com
or
Bonaire Partners, website: www.bonairepartners.com
Tel: +599 717 4545;
fax: +599 717 4544.
E-mail: bert@bonairepartners.com

NETHERLANDS ANTILLES - CURACAO (PJ2)
Caribbean Contesting Consortium's PJ2T contest station, Signal Point (IOTA SA-006)
Website: www.pj2t.org

Signal Point is presently available for rental to hams and their families and friends only. Members of the 'Caribbean Contesting Consortium' (CCC), who own the house and station at Signal Point, get first priority and receive preferred pricing because they contribute to the monthly operating costs.

The house consists of two air-conditioned bedrooms with two twin beds each, a bathroom each, and a common room which also includes the kitchen and the shack. There's a sizeable library of books. An additional room, the 'East Sunroom', is still under construction. There are two large covered outdoor porches facing the ocean and a covered eating area at the edge of the ocean cliff. The home is clean, but not luxurious, in order to keep costs under control. There is no Jacuzzi, hot tub or pool, but the house is on the ocean side, atop a beautiful coral cliff overlooking the Caribbean.

In contrast to the utilitarian house, the amateur station is of world-class

Stacked towers at the PJ2T station.

Another view of the world-class antennas at the PJ2T station.

PJ2T is located on top of a cliff, overlooking the Caribbean.

calibre. There are four HF stations plus an IC-505 6m multi-mode transceiver and a 2m FM rig. Station 1 is a Yaesu FT-1000MP and Ameritron AL-1200 linear; Station 2 an FT-1000MP MkV Field and Ten-Tec Titan II linear; Station 3 a computer-controlled Ten-Tec Omni VI transceiver and Amp Supply LK-800 linear and Station 4 a Kenwood TS-930 and Alpha 76 linear. Each station has Pentium desktop computers with Ethernet network, Bencher paddles etc. The Internet is available at Stations 1 and 3.

Outside are three towers: the 'Europe Tower' is 100ft and sports a 2-element 40m Yagi at 105ft (rotatable); 5-elements on 20m at 95ft fixed on Europe; 5-elements on 15m at 85ft fixed on Europe; 5-elements on 10m at 75ft fixed on Europe; and a Cushcraft A3S tribander at 45ft fixed on South America.

The 'US / JA Tower' is: 80ft high and supports a 5-element 6m Yagi at 85ft; a 3-element CL-33 tribander at 85ft (rotatable); a 5-element 205CAS at 67ft fixed on USA / JA; an 80m inverted-Vee; an XM-510 at 58ft fixed on USA / JA; and an XM-510 at 36ft fixed on USA / JA.

The 'WARC Tower' is 54ft and has a Cushcraft A3WS beam for 12 and 17m and a 30m Inverted-V at 50ft.

Other antennas include a 160m inverted-Vee at 100ft; a 3-element 80m delta loop array at 80ft favouring Europe, and a 540ft USA / JA Beverage low-band receiving antenna.

The CCC's philosophy is to make Signal Point as accessible to the ham community as possible, and at the lowest possible cost. CCC is a non-profit operation, and is renting at a price that

A room with a view.

Geoff, W0CG, operating as PJ2DX from Signal Point.

is set only to partially cover costs. Your rental helps CCC to keep the facility maintained and on the air.

The house is available in 'Bed and Breakfast' mode at certain times of the year when you will have full use of the station, one bedroom and one bathroom, and breakfast will be served daily. The owners will be in the other bedroom.

The rent is USD $800 per week for one or two persons 'Bed and Breakfast', or $1300 per week for one or two persons ($1450 for three or more persons) for the use of the full house, plus surcharges. The additional charges are a Utilities Surcharge, Station Support Contribution (used to support the operation, maintenance, and improvement of the station), and an Outside Mainte-

nance Surcharge (for tower maintenance). For full details please see the website or contact the owners.

Rental of Signal Point must be for at least one week, normally Tuesday to Monday (in order to accommodate contest weekends). Additional days after the first week are no problem.

Contact: Geoff Howard, W0CG, 1984 Trares Rd, Suffield, OH 44260, USA.
E-mail: ghoward@kent.edu

NEW ZEALAND

Quartz Hill, ZL6QH station, near Wellington (IOTA OC-036)
Website: www.zl6qh.com
The Quartz Hill amateur radio station, ZL6QH, was dismantled in October / November 2007 to make way for the

Quartz Hill, ZL6QH

Close-up of two main towers at the PJ2T superstation.

building of an electricity wind farm on the site. Brian Miller, ZL1AZE, Chair of the Quartz Hill Committee, Wellington Amateur Radio Club Inc, said in 2008, *"We do plan to rebuild the station but it could take up to two years or longer before it is available again for normal operations."* Any change in this status will be published on the group's website.

Until it was dismantled, ZL6QH was arguably the best HF DX facility in the South Pacific. The location was home to Radio New Zealand's MF and HF receiving station until 1997, when it was leased by the Wellington Amateur Radio Club. Members of the ZL6QH group converted the existing 20m wooden poles and 41m-high mast into an amateur radio antenna farm including 300m-long Vee-beams, a rhombic, monoband Yagis, and various wire antennas for the low bands.

The facilities at Quartz Hill were open to all members of the Quartz Hill Supporters' Group. Until the station was closed at the end of 2007, the annual subscription was NZD $50 per year (approximately GBP £20).

Contact: Chair of Quartz Hill Committee, Wellington Amateur Radio Club, PO Box 6464, Wellington 6141, New Zealand.

E-mail: chairperson@ZL6QH.com

NICARAGUA

YN2N guest house, Mombacho Volcano, near Granada
Website: www.yn2n.com

The YN2N Guest House.

The YN2N Guest House is located in a rural area just 10 minutes from the colonial city of Granada and 52km from Managua on the west coast of Lake Cocibolca. Granada is the oldest city in

Tower at YN2N.

Latin America founded by the Spanish and is a major tourist destination in itself. All the facilities you would expect in a city, including supermarkets, banks, hospitals, cinemas and good nightlife can be found in Granada.

At the guest house are two bedrooms with air conditioning and hot water showers. Visitors also enjoy the best Nicaraguan dishes made by Mrs Martha de Miranda, famous in Granada for her desserts. Octavio commented, *"My wife Martha and I are beginning with this project and I think it will be successful because we are trying to give the best of ourselves."*

Using the YN2N Guest House means you can avoid the cost and hassle of importing equipment and have less luggage to carry. All the equipment and antennas are set up and ready for use. The shack is comfortable and has a panoramic view with no background noise. The equipment is an Icom IC-706 transceiver and there are plans to buy a new Yaesu FT-450 transceiver. There are two linear amplifiers, a Drake L4 and a Heathkit SB-200. Antennas are a Cushcraft A3S 3-element Yagi for 10, 15 and 20m, a 2-element Moxon beam for 17m, full-size dipoles for 160,

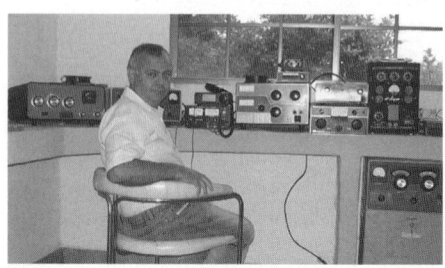

Octavio, YN2N, at his station.

Cushcraft A3S tribander and sloping dipoles at YN2N.

80 and 40m, a sloping dipole for 60m and a 3-element quad for 6m.

Airport pick-ups can be arranged, as well as transport to any of the local sightseeing destinations such as the Mombacho Volcano Reserve, Apoyo Lagoon, Colon Central Park, San Francisco Museum and a boat tour of Granada's 'Little Islands'.

Contact: Octavio Miranda, YN2N / TI2OHL, PO Box 257, San Jose 1017, Costa Rica.

Tel: +505 855 0852.

E-mail: omirandalac@yahoo.com

NORWAY

(SEE ALSO SVALBARD)
ARIM, Morokulien, LG5LG / SJ9WL station
Website:
www.east.no/priv/la7tia/arim
Morokulien is a 'ham state' on the Norwegian-Swedish border, and 2008 marked the 40th anniversary of the agreement by radio amateurs from Norway and Sweden to cooperate in running an amateur radio station on the border between the two countries. The location is unique, because it is prob-

ably the only amateur radio station in the world which is located right on the border between two countries.

The name Morokulien is a compound word from the Norwegian "*moro*" and Swedish "*kul*", both meaning "fun". ARIM, standing for 'Amateur Radio in Morokulien', is the foundation that maintains and runs the station. Its main aim is to provide funding for courses to enable disabled people to take out ama-

VHF/UHF and HF log-periodic Yagis at Morokulien.

The LG5LG/ SJ9WL tower and antennas.

Assembling the new HF log-periodic Yagi.

teur radio licences.

Morokulien is located near the Swedish town of Charlottenberg. From the Swedish side of the border follow main road 61. From the Norwegian side, follow main road 2 from Kongsvinger. Trains from both Oslo and Stockholm stop at Charlottenberg.

The station is located in the Border Cottage and the national border runs through the building. The cottage is wheelchair friendly: all doorsteps except at the entrance door have been removed and there is wheelchair access from the car park up to the entrance porch. All the doors are wide enough for wheelchairs.

There is a double bunk bed, plus two double sleeping sofas in the living room (bring your own linen or sleeping bags). The bathroom is large with arm rests on the toilet for wheelchair users. There is a small kitchen with refrigerator, microwave oven and coffee machine. The heating system in the cottage is electric but you can also make a log fire if you wish.

The radio shack is very popular because of its attractive callsigns: the Norwegian LG5LG and the Swedish SJ9WL. If you are visiting Morokulien for several days you should change the callsign you use from LG5LG to SJ9WL and vice versa every other day.

The shack is also wheelchair-friendly. All the equipment is from Icom, and is in good condition. The IC-765 has been in service for over 10 years, and replaced an earlier IC-720A which had made close to 1 million QSOs! An IC-706MkIIG is used for 6m, 2m and 70cm, as well as a back-up for HF if required. (Note that Sweden has restrictions on the use of 6m, but the band can be used with the Norwegian LG5LG callsign.) The shack also has a rare piece of equipment, a Siemens 100 radio teletype printer in good condition,

The granite peace monument on the Norwegian-Swedish border.

Close-up of the border cottage at Morokulien.

which has been equipped with a modem and built into a sound-proof console.

During the summer of 2006 a new mast and antennas were put up at Morokulien. The mast is 32 metres (105ft) high and sports an 8-element log-periodic Yagi for 10, 12, 15, 17 and 20m, with dipoles for 30, 40, 80 and 160m. For VHF - UHF there is a 21-element log-periodic antenna for 80 - 1680MHz, a 6-element Yagi for 6m, and a triband vertical for 2m, 70cm and 23cm.

The rent for one to five people is SEK / NOK 300kr (approx GBP £23 or £27 respectively) per night. If there are six people or more, the price is doubled. You pay when you pick up the keys at the customs station at the border. The money is used for the upkeep of the building and maintenance of the station and all profits go to the ARIM Foundation, 50% to Norway and 50% to Sweden.

If you wish to use the Morokulien station, please contact the booking manager, Odd, SM4SXQ.

Contact: Odd Westby, SM4SXQ, Glasbruksvägen 27, SE-673 92 Charlottenberg, Sweden.

E-mail: odd.westby@telia.com

PALAU

Palau Pacific Resort Hotel (Room 214) (IOTA OC-009)
Website: www.api-japan.com/palau/radio/english/radio1-e.htm (equipment); http://palauppr.com (hotel)
The Japanese Api Corporation runs two guest shacks in the Oceania region, the other being at the Langkah Syabas Beach Resort near Kota Kinabalu in Sabah, East Malaysia (see page 97).

The Palau Pacific Resort is a five-star luxury hotel. It overlooks the sea and nearby islands yet is just 20 minutes from the international airport. Room 214 at the resort is fully equipped with Yaesu amateur radio equipment including a new FT-2000 transceiver, an FT-1000MP with 500Hz CW filter, an FT-920 HF / 6m transceiver, and a Quadra VL-1000 1kW linear amplifier. Ancillary equipment such as a 12V PSU, rotator, antenna switch etc are also provided.

The antennas are set up on a 14m tower on the top of a hill and include a Force 12 C3 beam for 10, 15 and 20m,

The Palau Pacific Resort Hotel.

Antennas at the five-star Palau Pacific Resort Hotel.

a rotary dipole for 12, 17 and 30m, dipoles for 80 and 160m and an 8-element 6m Yagi, as well as a ground-mounted Force12 EF-140FV 40m full-size quarter-wave vertical.

The hotel can arrange tours to local attractions including the 300 mushroom-shaped 'rock islands', a dolphin research facility, the Badrulchau stone monoliths, World War II relics, jungle waterfalls, or just a tour of Koror city.
Contact: Ms Hiroko Tani at Api Corporation, Japan.
E-mail: apiapi@dream.com

VIP Guest Hotel, Koror (Room 303) (IOTA OC-009)

The VIP Guest Hotel is a small private hotel located at downtown Koror and within walking distance of all government offices, shopping malls and restaurants. The station is located in Room 303 on the third floor. There are three transceivers available, a Yaesu FT-767GX, a Kenwood TS-450S and a TS-850S, with a Kenwood TL-933 linear amplifier. The antennas are on the roof, 65ft high, and are a Create CL20 4-element Yagi and a Cushcraft R7 vertical covering 10, 12, 15, 17, 20, 30 and 40m.

The room rate for single or double occupancy is USD $60.50 per day, including tax. Just e-mail to reserve the room.
Contact: George Ngirarsaol, T88GN, VIP Guest Hotel, PO Box 18, Koror, Palau.
Tel: +680 488 4618 *or* +680 488 1502.
E-mail: vipghotel@palaunet.com

PARAGUAY

ZP5AZL station, Asunción
Website: www.qsl.net/cx6vm/ZP5AZL

You can enjoy being on the other side of the pile-up by visiting the ZP5AZL station in Asunción, the capital city of Paraguay. Asunción has good restaurants, good shopping and many beautiful colonial style buildings. Elsewhere in Paraguay are the Iguacu Falls, one of the natural wonders of the world, about 300km from the capital.

The ZP5AZL set-up is a world-class contest station and has been heard as ZP0R and a number of other special ZP contest calls over the years. The shack is air conditioned and is equipped for SO2R operation with one Icom IC-756 and two IC-765 transceivers, a Henry 3K-Ultra, a Drake L7B and a Heathkit SB-200 linear amplifier, two Pentium PCs, a WX0B 'Six Pack' antenna selection system, Daiwa 12V PSU etc.

VIP Guest Hotel in Koror, Palau.

The ZP5AZL house from front (left) and rear (right).

There are five towers with a 6-element monoband Yagi for 10m, a 5-over-5-over-5 array for 15m, a 4-element monobander for 20m, two Hy-Gain TH6DXX beams for 10, 15 and 20m, a 2-element Yagi for 40m, a 3-element Yagi for 12 and 17m, and dipoles for 30, 80 and 160m.

The cost of staying for one week, inclusive of all meals (breakfast, lunch and dinner) and use of the radio is USD $800 for one person or $1000 for two people. Additional guests can stay at a hotel nearby for a very reasonable price. The rates include transport to and from the airport.

Contact: Tomàs (Tom) Anibal Zapattini, ZP5AZL, Av Mariscal López c/ 26 de Febrero, Asunción, Paraguay, *or* Senador Long 370 C/ Andrade, Asunción, Paraguay.
Tel: +595 21 603935.
E-mail: azl@quanta.com.py

PHILIPPINES

Boracay DX Society, Boracay, Visayan Islands (IOTA OC-129)
Website: http://members.tripod. com/boracaydave/mycompany.html
The Boracay DX Society has a five bedroom, three bathroom local style house built on four levels. The location is between Nami Resort and Microtell Resort on Boracay's Din-Iwid Beach, the Philippines' premier tourist resort. The property is on a 200ft cliff overlooking the beach with views to six other islands and spectacular sunsets. There are terraced gardens with fruit trees, orchids and a waterfall.

The station runs full legal power on all bands and includes a Kenwood

The beach at Boracay.

6-element monobander on 10m, 4-element monobander on 20m.

TS-50S and TS-440, an Icom IC-706, a Yaesu FT-747, and vintage HW-9, Viking Adventurer and SX-53S rigs. Amplifiers are a home-brew 4CX1500, a Dentron GLA-1000B and home-brew 813s. Antennas are a Cushcraft A3S 3-element Yagi for 10, 15 and 20m and a G5RV antenna.

Clubs and DXers are invited to visit and reserve the facilities. The rates, which include the use of a car and boat, are USD $50 per day or $100 per week, and donations are accepted in lieu of payment.

Contact: Dave or Lynn Wilson, Boracay DX Society, c/o NVC, Tirol Center, Main Road Mangayad, Boracay, Malay, Aklan, PI 5608, Philippines.
Tel: +63 36 288 3873.
E-mail: boracaydave@yahoo.com

PORTUGAL

CT1IUW self-catering cottage, near Praia da Luz, Algarve
Website: www.ownersdirect.com (reference P2758)
Don Stewart, G3TIR / CT1IUW, and his wife Mavis retired to a villa in the countryside near Lagos, Algarve, in 2003. They offer economical accommodation in a separate self-catering apartment sleeping four, with the use of the shack in the villa.

Set within the grounds of the villa, the cottage has a master bedroom with double bed overlooking the garden and with a door leading to a raised decking area; bedroom 2 with twin beds; bathroom; lounge / diner with satellite TV, CD player etc; and kitchen with cooker, hob, oven, fridge / freezer, microwave oven and a door leading to raised decking overlooking the garden with table, chairs and sunshade for *al fresco* dining. Also outside is a good-sized private swimming pool.

The station (in the villa) comprises a Kenwood TS-930S to an 80m inverted-Vee tuned doublet antenna. Broadband Internet access is also available.

There are several golf courses close by, including a new 18-hole course within walking distance of the cottage, which is scheduled to open during 2008.

Espiche, half a mile away, is a quaint village with small bars, two ex-

CT1IUW's villa.

PHOTO: ALISON STEWART

cellent restaurants and a small super-market. Praia da Luz is 1.5 miles from the cottage and has a sandy beach, exclusive shops and boutiques, a good variety of cafes, restaurants and bars, live music, a night club, cinema, supermarket, post office and Internet cafes. Three miles away is the town of Lagos with numerous historic sights as well as a pedestrianised centre with bars, cafes and restaurants and a wide variety of sandy beaches and coves. Sagres, mainland Europe's most westerly point, is an easy drive away, and leads to the vast empty beaches of the west coast.

The cottage is just over an hour's

Kitchen at the CT1IUW self-catering cottage.

drive from Faro international airport. Car hire is recommended, although the local taxi service is very good.
Contact: Don Stewart, CT1IUW, Apartado 347, 8601-929 Luz / Lagos, Portugal.
Tel: +351 282 788 036.
E-mail: donstime@sapo.pt
or
UK agent: Miss Alison Stewart,
Tel: +44 789 467 0235.
E-mail: stewarta@fsmail.net

SAN MARINO

T70A club station
Website: www.arrsm.org
It is not possible for foreigners to receive an HF visitor's licence in San Marino (although a VHF licence is possible: see the 'Licensing Information' section of this book for further details). However, subject to availability, the ARRSM club station, T70A, is open to foreign visitors. A 7-element log-periodic beam and a 3-ele tribander are available but you must use your own rig and log all contacts using an approved computer logging program so that ARRSM can respond easily to QSL requests.

To request permission to use the station, write to the T70A Radio Club at least two months before the expected operation date.
Contact: ARRSM Radio Club, PO Box 77, 47890 San Marino A-1, Republic of San Marino, Italy.
E-mail: t70a@arrsm.org

SCOTLAND

Dunnet Head bed & breakfast and self-catering accommodation, near Thurso, Caithness (IOTA EU-005)
Website: www.dunnethead.co.uk
Dunnet Head is mainland Britain's most northerly point - north of Moscow or Stavanger in Norway! Brian Sparks, GM4JYB, offers both self-catering and bed and breakfast accommodation with generous discounts for radio amateurs.

Use can be made of Brian's station, consisting of Yaesu FT-1000MP and Icom IC-7000 transceivers covering all bands from 160m to 70cm at a maximum of 100W output, with computers available for the digital modes. The main HF antenna is a Cushcraft A3S 3-element Yagi on a 30ft telescopic mast. As for the other antennas, Brian says, *"It depends on what the wind leaves me with at the end of the winter. Hopefully for 80, 40 and 30m it will be dipoles up at about 40ft along the top of a cliff overlooking the Pentland Firth and possibly a vertical for 160m. The Cushcraft A3S has been put away for the winter so should be OK! I like operating on 17m and will be looking to make a 'Hexbeam'."*

The property overlooks the Pentland Firth and rock stacks in Brough Bay. The area is superb for walking, cycling and wildlife enthusiasts: puffins visit the nest in their burrows on Dunnet Head from May to July, whales are often seen and there is a resident seal colony in Brough Bay behind the property. Visitors can enjoy a wildlife walk with the Highland Ranger, Forestry Commission or RSPB, cycle in the nearby forests or on the quiet roads, or fish in the Dunnet Head lochs (permits available for purchase). A qualified guide is available for walking tours and tours of the WWII radar installations on Dunnet Head. There is good surfing at Dunnet Bay, three miles away, and facilities for kayaking in Brough Bay and on the Dunnet Head lochs. The nearest golf courses are at Thurso (14 miles) Wick (20 miles) and Reay (21 miles).

The self-catering accommodation consists of two double bedrooms, kitchen, lounge and access to the garden which overlooks Pentland Firth, rock stacks and the seal colony. They come fully equipped with linen, crockery and cutlery, washing machine, oven, hob, microwave and TV. The accommodation is available from Easter until October each year and the rates (inclusive of heating and electricity) are: March to June £400, July to August £450, September to October £400 per week (normally Saturday to Saturday but other periods might be possible). There is a £50 per week discount for radio amateurs.

For those not wishing to stay as long, bed and breakfast accommodation is also offered with prices from £20 per person per night for double occupancy or £26 per night for single occupancy. Discounts are available for stays of more than three nights.

Contact: Brian Sparks, GM4JYB, Windhaven, Brough, Thurso, Caithness, Scotland, UK.
Tel: +44 1847 851774.
E-mail: gm4jyb@dunnethead.co.uk

SENEGAL

Le Calao, Somone
Website: www.le-calao.com
Le Calao is a small French-owned private resort in the village of Somone, about 70km (two hours by car) southeast of Dakar. The resort is between the Atlantic Ocean (500m to the west) and a salt-water lagoon (300m to the north), and is a stone's throw from the centre of the village with its shops, restaurants, etc.

There are four furnished bungalows for rent, each designed for up to two adults and a child. The bungalows are set in a garden with trees and tropical plants and are adjacent to the swim-

Titanex V80 vertical in the salt marsh.

ming-pool. Each bungalow is fully furnished and includes a kitchen with a gas hob, a sink and a large fridge. In the bedroom there is a ceiling fan and a movable standing fan.

Mountain bikes can be rented for rides in the local area or to go shopping in the village.

The shack at *Le Calao* is located on the first floor of a separate building. The operating room has two fans and can accommodate four or even more people. A large fitted out operating bench allows three stations to be set up in no time. The shack has been specially designed to allow for quick modification, such as connecting up an additional antenna.

The following equipment is available: a Kenwood TS-870S HF transceiver, and a Kenwood TM-133 2m / 70cm rig allowing for SSB, CW, FM, RTTY, PSK, Packet, APRS and *Echolink* etc. There are several 20A and 32A power supplies on a shelf under the operating bench. It is recommended that you take your own headphones, keyer and PC for logging or other software. A 95Ah battery is provided in case of interruptions to the mains electricity supply (which happens quite often at some times of the year).

Free 512K ADSL Internet access is available 24 hours a day via four Ethernet RJ45 ports or via wi-fi.

There is no linear amplifier provided. Guests may take their own and use it as long as it does not cause interference to the TVs, hi-fi, wi-fi etc.

There are 3-element monoband Yagis for 10m and 20m on a 45ft tower and another 45ft tower supports a 3-element monoband Yagi for 15m as well as a VHF / UHF collinear vertical. In the field towards the salt-marsh lagoon and 150ft from the shack is a Titanex V80 vertical which is used on 12, 17, 30, 40, 80 and 160m via its base matching unit. A Tonna 5-element Yagi for 6m is located above the shack roof. It is easy to put up more antennas if required.

The costs are as follows. Shack and lodging for two or three people EUR 550 euros (approx GBP £393) for a full week. For a special contest period (three days from Friday 12.00am until Sunday 12.00am) the shack only for one opera-

Senegal: Le Calao.

tor (without lodging) is 200 euros (approx £143), or shack and lodging for two or three people 300 euros (approx £215). For the use of the shack only for one person (without lodging) 180 euros (approx £129) for the week-end or 400 euros (approx £286) for a full week. If you take a linear amplifier there is an extra charge for electricity of 2 euros (approx £1.40) per day up to 600W output, or 4 euros (approx £2.85) per day for 600 - 1500W output.

Transport from the airport is 40 euros (approx £29).

Contact: Jan-François Lorne, 6W7RV/ F4AHV, BP 06 Ngaparou (Mbour), Somone, Senegal.
Tel: +221 339 585 131;
mob: +221 775 688 018.
E-mail: info@le-calao.com *or:*
Jan-François Lorne, F4AHV, 3 rue Louis Rolland, 92120 Montrouge, France.
Mob: +33 60 922 1009.

The well designed shack at Le Calao.

NOTE: A DXPEDITION ON YOUR DXPEDITION?

Jan-François, 6W7RV, can help you to organise a portable DXpedition in a 4X4 vehicle equipped with radio from 80m to 70cm. The vehicle has two 105Ah batteries with 400 / 800W 12V - 220V inverter and is equipped with an SGC SG-230 SmartTuner for 1.8 - 30MHz (max 200W) with SG-202 antenna, and Maldol HFC-40L, HFC-30L and HFC-20L mobile antennas for 40, 30 and 20m. The 4X4 vehicle has a special 50 litre water tank and a 45 litre fridge.

Suggested destinations are the Republic of Guinea (3X), Mauritania (5T), the Gambia (C5), Guinea-Bissau (J5) and Mali (TZ). Contact Jan-François for further details: Jan-François Lorne, 6W7RV/ F4AHV, BP 06 Ngaparou (Mbour), Somone, Senegal.
Tel: +221 339 585 131;
mob: +221 775 688 018.
E-mail: info@le-calao.com *or:*
Jan-François Lorne, F4AHV, 3 rue Louis Rolland, 92120 Montrouge, France.
Mob: +33 60 922 1009.

SOUTH AFRICA

African DX Safaris
Websites: www.dxsafari.com *and* **www.africandxsafari.com**
African DX Safaris offer DXpeditions in South Africa, Botswana, Swaziland, Lesotho and Mozambique. They can either be 'pure' DXpeditions, or associated with traditional game-viewing safaris. The trips are tailor-made for each group and anyone wanting to go on a 'DX Safari' can request that one be or-

African DX Safaris' trailer tower.

ganised at any time.

At present there are two Yaesu FT-1000s and Icom IC-2KL linear amplifiers available, along with two 50ft trailer-mounted crank-up towers. The antennas are two Tennadyne T8 log-periodic Yagis, a 40m 4-square array, a 70ft vertical that works on 160, 80 and 30m and an M^2 6m beam. One location used, in Swaziland, also has a 60ft tower with Cushcraft A4S and A3WS beams for all five bands from 10 to 20m.

African DX Safaris can arrange for game safaris from two to five days in duration, with or without amateur radio. All anyone needs to take is their paddle for CW and a laptop for logging.
Contact: USA representative Charles 'Frosty' Frost, K5LBU, 3311 Hilton Head Ct, Missouri City, Texas 77459, USA.
Tel: +1 281 682 6093.
E-mail: frosty1@pdq.net *or*
Andre van Wyke, KR5DX / ZS6WPX
E-mail: kr5dx@yahoo.com

SOUTH COOK ISLANDS

Gina's Garden Lodges, Aitutaki (IOTA OC-083). Websites:
www.ginasaitutakidesire.com
www.cookpages.com/GinasLodges-Aitutaki *and* **www.ginasaitutaki.com** (three different websites)
Gina's Aitutaki Island Holidays offers accommodation in two locations at

Two of the four lodges, with pool and sun decking.

One of Gina's Garden Lodges, with antenna at left.

Inside Gina's Garden Lodges.

Aitutaki. One is Gina's Garden Lodges at Tautu Village on the main island of Aitutaki, and the other Gina's Beach Lodge on the uninhabited *motu* (islet) of Akaiami in the Aitutaki lagoon. Both are run by Paramount Chief Queen Manarangi Tutai and her husband Des Clarke, E51DD.

Des writes, *"We are an amateur radio friendly establishment and have in the past had amateur radio antennas at our garden lodges for guest use, although we have never had station equipment for rental. However, successive hurricanes have pretty much eliminated the antenna farm, leaving only one home-brew 3-element monoband Yagi antenna for 20m. We will always welcome hams to stay with us, and do our best to assist them, but they should bring their own equipment if they wish to operate."*

The 20m monobander is available for the use of guests staying at Gina's Garden Lodges. There are four lodges with self-catering facilities set in a two acre garden with swimming pool and sun deck. Each lodge is complete with cooking and tea and coffee making facilities, refrigerator and oscillating fans.

Gina's Beach Lodge is situated on a pristine white sandy beach five miles across the lagoon on Akaiami island. Here, there are three studio-type rooms with colonial-style verandah and separate cooking, toilet and shower facilities. Electricity is provided. The Beach Lodge provides doorstep access to excellent swimming and snorkelling. No amateur radio equipment or antennas are provided at the Beach Lodge.

Arrangements can be made for 'island nights' at local restaurants, lagoon tours, diving excursions and hire vehicles if required.

The rates at Gina's Garden Lodges are NZD $120 (approx GBP £48.50) per night for a double/twin lodge or NZ $75 (approx £30) per night for single occupancy. Guests who book for seven nights pay for six (one night free

Gina's Beach Lodge is located on the motu of Akaiami.

Akaiami motu across the lagoon.

of charge). The rates for Gina's Beach Lodge at Akaiami (including boat transfers across the lagoon) are NZ $300 (approx £121) per night for double/twin or NZ $180 (approx £73) per night for a single. Guests are met at Aitutaki airport and transfers are NZ $16.

Contact: Des Clarke, E51DD, Gina's Garden Lodges, PO Box 10, Tautu Village, Aitutaki, Cook Islands.
Tel / fax: +682 31058.
E-mail: queen@aitutaki.net.ck

SPAIN

EA5ON villa, nr Valencia
Website: www.homelidays.com/ la-eliana/house-villa180929en1.htm
Duncan Lindsay, EA5ON / GM7CXM, offers a three-bedroom detached villa with living room, kitchen, bathroom and, outside, a small garden, swimming pool and barbecue. There's parking for up to three cars.

There is a 40ft guyed tower with a Mosley TA-53M beam for 10, 12, 15, 17 and 20m and in the shack a Kenwood TS-850S (with Inrad SSB filters and a voice keyer). There are no low-band antennas, although a 40m dipole will fit in the garden: 80 and 160m would be more complicated, but are 'doable'.

The house is located in a nice town,

EA5ON's villa, nr Valencia.

Antennas at the PZ5RA station.

Pool at the EA5ON villa, with tower at left.

15km from Valencia city centre and close to the airport (approx 20 euro for a taxi). It is a 10-minute walk to the town centre or half an hour to Valencia by train or bus. Valencia offers extensive beaches, an historic city centre, opera house and a wonderful science city designed by Santiago Calatrava.

Duncan says that *"this is not primarily a DX rental since ours is not what you might call a rare DX location. However, for the time being the tower remains, with a Mosley TA-53M on top, and a 100W station is available to anyone who might wish it."* The villa is generally rented out as a holiday home to non-radio amateurs on a weekly basis. The rent is from EUR 600 to 800 euros per week.

Contact: Duncan Lindsay, EA5ON / GM7CXM, Avda de las Delicias 41, 46183 La Eliana (Valencia), Spain.
Tel: +34 96 335 9107;
fax: +34 96 335 9101.
E-mail: ea5on@ea5ol.net *or* dlindsay.vlc@mscspain.com

SURINAME

PZ5RA station, Paramaribo
Website: www.qrz.com/pz5ra
If you would like to have huge pile-ups or if you are a contester and you would like to visit Suriname, Ramon, PZ5RA, is now offering accommodation and the use of his station in Paramaribo.

The station consists of the following equipment: Yaesu FT-920 and Kenwood TS-50 transceivers to Acom 2000A and Kenwood TL-920A linear amplifiers. The antennas are a new Tennadyne T10.10-30HD 10-element log periodic for 10 - 30m (installed in November 2007), a Mosley PRO-67B 7-element beam for 10m to 40m, an Alpha Delta DX Sloper for 30, 80 and 160m, a dipole for 80m and a home-made 6-element Yagi for 6m. All digital modes can be used via a sound card using *Mixwave* software.

Contact: Ramon Kaersenhout, PZ5RA, PO Box 745, Paramaribo, Suriname.
Tel: +597 490 976 *or* +597 088 37720.
E-mail: pz5ra@amsat.org *or* pz5ra@hotmail.com

SVALBARD

JW5E club station, Longyearbyen, Spitsbergen (IOTA EU-026)
Website: http://home.online.no/ ~polar-ps/frametest.htm
The NRRL Svalbard Group has made its club station, JW5E, at Longyearbyen available for visitors. The club cottage is about 45 sq m in size: the radio shack is 15 sq m, with 30 sq m for relaxing, sleeping and cooking. You can make your own food and sleep at the club station (take a sleeping bag) but note that there is no running water or bathroom at the shack. Members of the club will bring you fresh water, while bathrooms, showers and a sauna are at the hotel about 700 - 800 metres from the club building.

JW5E club station in centre of picture.

5-element Fritzel beam at JW5E.

Equipment includes a complete Icom line provided by the LA-DX Group, comprising of an IC-751 transceiver, an IC-2KL linear amplifier and an ATU, and the antennas are a 5-element Fritzel beam and 40 and 80m dipoles on a 100ft tower. You can either use your own callsign with the JW prefix or the club call, JW5E. The maximum number of operators is three, or four to five for a 48-hour contest with one or two operators on the air at any time.

The rate charged is NOK 400 kroner or EUR 50 euros per day (approx GBP £38) with a special price for LA-DX Group members of 300 kroner (approx £29) per day. The sum should be paid to the club's bank account or by *Paypal* before arrival in Svalbard. Further details from Math, JW5NM.
Contact: Mathias Bjerrang, JW5NM, Svalbard Lufthavn, Postboks 498, 9171 Longyearbyen, Norway.
E-mail: polar-ps@online.no

JW5E antennas.

SWAZILAND

African DX Safaris
African DX Safaris can organise tailor-made DXpeditions to Swaziland, among

PHOTO: 9M6XRO

Fixed tower at African DX Safaris' Swaziland location.

PHOTO: 9M6XRO

other countries, providing accommodation, meals, equipment and antennas.

In addition to the two Tennadyne T8 log-periodic Yagis and other antennas transported with the trailer-towers, the location used by African DX Safaris in Swaziland has a fixed tower with a Cushcraft A4S beam for 10, 15 and 20m

at 79ft and an A3WS beam for 12 and 17m at 66ft, as well as a 4-Square for 40m and a 72ft vertical for 80 and 160m that can be raised by two people when required.

For contact details, see the African DX Safaris' main listing under SOUTH AFRICA.

DX from the high veld: Chalet 19 houses the shack, with the tower and beams behind.

PHOTO: 9M6XRO

Swaziland: 72ft vertical.

SWEDEN

King Chulalongkorn Memorial Amateur Radio Society, SI9AM, Ragunda. Website: www.si9am.se
What is a Thai pavilion doing amid the pine and birch forests of central Sweden? In 1897, King Chulalongkorn of Siam visited Sweden and Ragunda municipality. One hundred years later, the construction of a memorial pavilion in his honour was begun. Today, the only Thai pavilion in the world outside Thailand (and the only one that is covered in a mantel of snow for a large part of the year!) stands at Utanede in Ragunda, at 62 degrees north latitude.

The visitor's amateur radio station

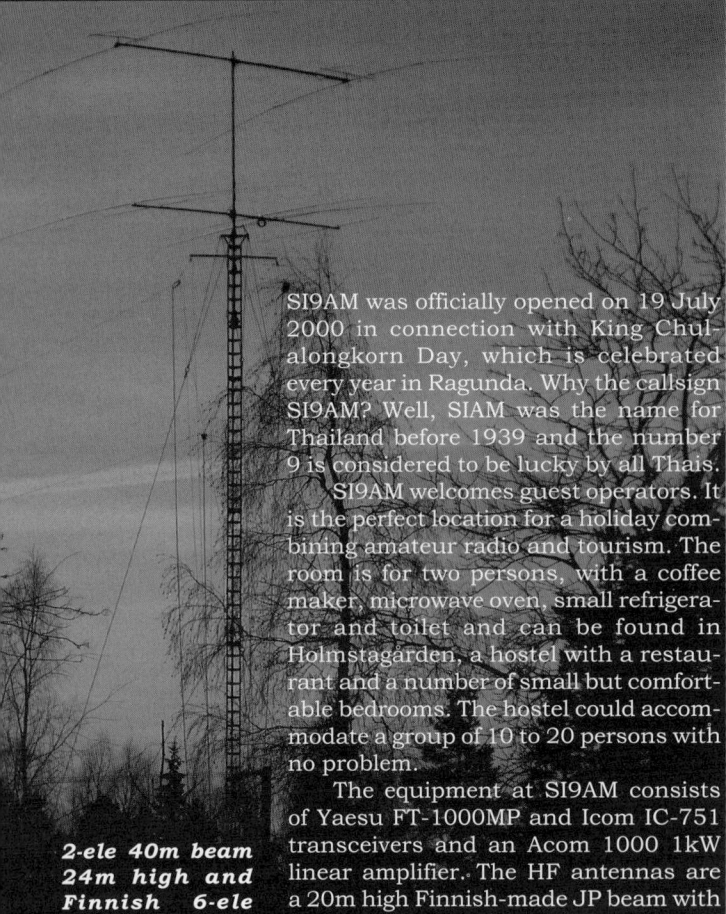

2-ele 40m beam 24m high and Finnish 6-ele triband beam at SI9AM.

SI9AM was officially opened on 19 July 2000 in connection with King Chulalongkorn Day, which is celebrated every year in Ragunda. Why the callsign SI9AM? Well, SIAM was the name for Thailand before 1939 and the number 9 is considered to be lucky by all Thais.

SI9AM welcomes guest operators. It is the perfect location for a holiday combining amateur radio and tourism. The room is for two persons, with a coffee maker, microwave oven, small refrigerator and toilet and can be found in Holmstagården, a hostel with a restaurant and a number of small but comfortable bedrooms. The hostel could accommodate a group of 10 to 20 persons with no problem.

The equipment at SI9AM consists of Yaesu FT-1000MP and Icom IC-751 transceivers and an Acom 1000 1kW linear amplifier. The HF antennas are a 20m high Finnish-made JP beam with six full-size trapless elements for 10, 15 and 20m; a 2-element beam for 40m (24 metres high); a folded dipole for 10, 12, 17, 20, 40 and 80m, and a 160m dipole. On 2m there is a Yaesu FT-225RD 25W multi-mode transceiver and an ARX-2 vertical antenna.

Aside from amateur radio, there is a number of interesting activities in the

Thailand? No, Sweden! The King Chulalongkorn memorial stupa at Ragunda, adjacent to the SI9AM guest shack.

area. The exotic Thai pavilion is only 200 metres away and offers guided tours every hour from June until August. During the winter group tours can be arranged but must be booked in advance. Utanede is a scenic location on the banks of the Indalsälven river with its fascinating 'dead waterfall', formed when Ragunda lake suddenly lost all its water in a period of under four hours. The rapids dried out and are now known as the dead falls. Hammarstrands radio museum is also in the area, as is the 'moosery' where you can see moose or elk in their natural environment.

During the summer period (1 May - 30 September) the daily rental fee for SI9AM is SEK 350 kronor (approx £27) for one or two people; 400 kronor (approx £31) for three people; 450 kronor (approx £35) for four people; or 500 kronor (approx £39) for five people. During the winter period (October to April) the daily rental is 50 kronor (approx £4) less in each case. For groups larger than five people, please contact Lars, SM3CVM, or Jörgen, SM3FJF, for

SM3FJF and SM3CVM at SI9AM.

a quote. Accommodation can be arranged in the same building as the SI9AM station at low cost.

Any changes, for example to the daily rates, will be published on the SI9AM website.

Contact: Lars Aronsson, SM3CVM, Lillfjällvägen 62, SE-831 71 Östersund, Sweden.
Tel: +46 63 850 09 or +46 70 343 0627.
E-mail: info@si9am.se

Kvarnberget club station, SK0UX, Vallentuna, near Stockholm
Website: www.sk0ux.se
Kvarnberget is a hill-top site located in Vallentuna, about 30km north of Stockholm. The site is operated by Kvarnbergets Amatörradioförening, SK0UX (special contest callsign SK0X).

The location was first used in 1956 for tropospheric experiments on 3000MHz by the Swedish Defence Research Institute (FOA). In 1978 FOA moved all its operations to Linköping, and Kvarnberget was closed down. Members of the Täby amateur radio club TSA (Täby Sändareamatörer) negotiated with FOA for the lease of the site and this was granted in 1982.

SK0UX's objective now is to provide members with a superb location and an excellent range of antennas from 1800kHz to 10GHz. To use the station

2-element cubical quad for 40m.

you should become a member of the SK0UX club. Rigs are not provided: simply take your own equipment and plug it in.

The antennas are being improved and are changing all the time, but the following is a representative sample of what is available. For 1.8MHz an inverted-Vee 35m high; for 3.5MHz a full-size 3-element Yagi on a 40m high mast (SK0UX's newest antenna); for 7MHz a 2-element monoband cubical quad on a 10m boom, mounted on a 35m tower; for 14, 18, 21, 24 and 28MHz a 12-element log-periodic, 24m high; for 21MHz a 4-element monoband Yagi at 15m high; for 28MHz an 8-element quad

Operating room and 10m 8-element quad at SK0UX.

Full-size 80m 3-element Yagi on 130ft tower.

on a 14.5m long boom, 18m high; for 7, 14, 21 and 28MHz a Fritzel FB-33 3-element tribander with 40m driven element at 12m high; for 50MHz a Comet 6-element Yagi; for 144MHz 2 x Cue Dee 15-element Yagis, 4 x M² 17-element Yagis and 2 x 16-element Yagis; for 432MHz 4 x Cue Dee 17-element Yagis and 2 x K1FO 22-element Yagis; for 1296MHz 2 x Tonna 55-element Yagis, a 2m dish and a 6m dish; for 2320MHz a 2m dish and a 6m dish; for 5760MHz a 90cm dish and for 10GHz a 90cm dish.

All members of the SK0UX club have access to the station. The membership fee is SEK 350 kronor (approx GBP £27) per year (less for under-20s).
Contact: Kvarnberget Amateur Radio Club, SK0UX, c/o Teemu S Korhonen, SM0W, Trädgårdsgatan 19 (8tr), SE-17238 Sundbyberg, Sweden.
Tel: +46 7060 30007 (President), +46 7024 36288 (Secretary).

ARIM, Morokulien (SJ9WL / LG5LG Station)
Website:
www.east.no/priv/la7tia/arim
See entry under NORWAY.

SWITZERLAND

See ITU Geneva.

SYRIA

Syrian Scientific Technical Amateur Radio Society (SSTARS) club station, YK0RJ, Damascus
Website: www.qsl.net/tir/Home.htm
Syria is not the easiest country in the world from which to attempt to operate amateur radio. However, the Syrian Scientific Technical Amateur Radio Society (SSTARS), which represents Syrian national radio amateurs, states on its website that the club station of SSTARS, YK0RJ, may be used by groups of foreign amateurs (not individuals) who are able to obtain an amateur radio licence in Syria (see the 'Licensing Information' section of this book for further details).

The club station is located at a Syrian Telecom Establishment (STE) building in Damascus. On a small tower

SSTARS club station, Damascus.

on the roof is a Force 12 C3 Yagi for 10, 15 and 20m, and a 2-element Yagi for 40m.

A UK group operated as YK9G from this station in April 2008.
Contact: Dr Omar Shabsigh, YK1AO, PO Box 245, Damascus, Syria.
Tel: +963 11 311 4540 *or* +963 11 612 1279 / +963 11 231 8796.
E-mail: shabs.om@scs-net.org

THAILAND

Radio Amateur Society of Thailand club station, HS0AC, near Bangkok
Websites: www.rast.or.th (mainly in Thai) *and* www.qsl.net/rast (English)
It is not very easy for non-resident foreigners to gain operating privileges in Thailand, but the Radio Amateur Society of Thailand, RAST (under the Royal Patronage of His Majesty the King) negotiated a special concession from the Thai Post Telegraph Department to allow suitably qualified foreign guests to operate the RAST club station, HS0AC. As such, the authorities (now represented by the National Telecommunications Commission) require that all guest operators be a member of RAST.

The RAST HS0AC club station.

Monoband Yagi at HS0AC.

This costs THB 2100 baht (approx GBP £35) for life membership. You must also have a valid licence in your home country, and bring a photocopy of this licence when you visit HS0AC.

The HS0AC station is located at the Asian Institute of Technology (AIT), Klong Luang, Pathumtani 12120, about 40km north of Bangkok on Paholyothin Road. Taxis from central Bangkok cost a minimum of 300 baht (approx £5), depending on traffic conditions at the time.

The station is not normally open or staffed, so arrangements for access have to be made through the Station Manager in advance. Access is subject to the availability of the Station Manager or his assistants.

Guest operators must use the callsign HS0AC, but may mention "operated by own callsign", or use "HS0AC/own callsign", e.g. HS0AC/M0QQQ (the former practice of using HS0/own callsign for guest operators has been discontinued).

There are photographs of the HS0AC station on the RAST Thai-language website at www.rast.or.th/hs0ac.html and full details of the rules for guest operation of the station can be found in English at www.qsl.net/rast/text/HS0ACrules.htm Contact the Station Manager, Finn, OZ1HET, for further details.

Contact: Finn Jensen, OZ1HET, 53/722 Krissadanakorn, Chaeng Watthana Road, Pak-Kret, Nonthaburi 11120, Thailand.
Mob: +66 6 533 7985
E-mail: oz1het@yahoo.com

HS1CHB guest shack, Bangkok
Major Narissara ('John') Shaowanasai, HS1CHB / N9WMS, rents out an apartment located on the top floor of the 14-storey Muang Thong Tanee exhibition centre. Licensed Thais and foreigners holding an HS0Z reciprocal licence may use the station at the apartment at no extra charge. It is equipped with a Yaesu FT-1000MP and a Kenwood TS-570D transceiver, with a 3-element beam for 10, 15 and 20m on the roof of the building about 50m (160ft) above the ground. A 40m dipole can be installed if required.

Note that with the single exception of the RAST HS0AC club station, non-Thai nationals must hold an HS0Z reciprocal licence in order to operate amateur radio in Thailand (see 'Licensing Information' section of this book for further details.)
Contact: Major Narissara Shaowanasai, HS1CHB, PO Box 1, Bangkok 10900, Thailand.
Tel: +66 2 589 2550;
mob: +66 81 1736467.
E-mail: n9wms@hotmail.com

TUNISIA

Association Tunisienne des Radioamateurs (ASTRA) club stations
Although it is not possible for foreign radio amateurs to gain their own licence and callsign in Tunisia, it is possible to operate the existing club stations in the country. Most of these have been set up for the Scouts in Tunisia by the Association Tunisienne des Radioamateurs (ASTRA) and with the cooperation of German radio amateurs. The club stations are:
3V8BB in Bir el-Bay, Tunis (at the *Institute Superieur de l'animation pour la Jeunesse et la Culture*);
3V8CB in Borj Sedria;
3V8SF in Sfax;

Scout club station 3V8SF at Sfax.

3V8SJ in Jendouba;

3V8SM in Houmt Souk, Jerba (Djerba, IOTA AF-083);

3V8SQ in Monastir;

3V8SS in Sousse;

3V8ST in Tunis and

3V8TZ in Tozeur (in the Tunisian Sahara).

The 3V8SF club station in Sfax has its own website at http://geocities.com/achraftn/3v8sf

For further details, contact the International Liaison Officer of the Association Tunisienne des Radio-amateurs (ASTRA), Mustapha, DL1BDF.

Mustapha is planning to set up three more stations in Tunisia at Bizerta in the north of the country, at Nabeul in the Cabon Region and in Kasserin.

Contact: Mustapha Landoulsi, DL1BDF, Westlinteler Weg 30, 26506 Norden, Germany.

Tel / fax: +49 4931 12519.

E-mail: info@landoulsi-norden.de

TURKS & CAICOS

The HAMlet, Providenciales Island (IOTA NA-002)
Website: www.vp5jm.com

The HAMlet is an airy island cottage set in nicely maintained grounds a few minutes walk from the beach on the island of Providenciales, or 'Provo'. The cottage has one comfortable bedroom with a queen-size bed and there is also

Sunset over the HAMlet antennas.

7-ele M² 6m Yagi on cottage's rooftop tower.

a fold-out couch in the living room. High-speed Internet is also provided for those bringing a computer with a network access card and there is cable TV in the living room. A large cushioned bench in the ham shack transforms into two independent single beds. The kitchen is fully fitted with refrigerator, gas stove, and microwave oven. The bathroom has a hot water shower.

The HAMlet has two operating consoles that can accommodate up to four separate stations to make anything from single-operator to multi-multi contesting a comfortable experience. Your can use the HAMlet HF transceiver with CW filter or you may choose to take your own equipment. No amplifier is available but there are two 220V AC outlets adjacent to each other for distribution to the various operating positions, plus several USA-style 120V AC outlets. The station includes an Astron 50M power supply that can provide 12V DC at up to 38 amps continuously, so your heavy 12V power supply can be left at home. There are a couple of ATUs that previous guests have generously left behind. You should take you own headphones, keyer and computer as well as band-pass filters if you intend to operate on more than one band at the same time.

The antennas include a Force 12 XR-5 multiband Yagi with a Force 12 Delta 240 above it, covering 40, 20, 17, 15, 12 and 10m. The HAMlet is located about 100ft above sea level on a moderate drop-off that allows these antennas on a 40ft self-supporting tower to perform like magic on the bands in all directions. A second tower has a 10m Yagi and a Force 12 75 / 80m dipole (both fixed). There is also a 53ft Hy-Gain AV-

18HT 'HyTower' all-band vertical. Various wire antennas complete the HF antenna system. For VHF, an M² 6M7JHV 7-element 6m Yagi on a 30ft boom is mounted on a rooftop tower on the cottage.

The rental fee is as follows: USD $170 for one person per night; $210 for two people per night; $1050 for one person per week; $1425 for two people per week and a $60 charge per night for each person over two people. Rentals that include a major contest must be for a minimum of one week. The HAMlet rental fee does not include air conditioning since many guests find that the normal tropical breezes aided by a fan or two provide a comfortable temperature. However, the bedroom does have a window-mounted air conditioning unit that can be activated for a minimal additional charge that covers the expensive cost of island energy.

Contact: Jody Millspaugh, VP5JM, PO Box 218, Providenciales, Turks & Caicos Islands, British West Indies *or* Jody Millspaugh, PO Box 694800, Miami, Florida 33269, USA
Tel: +1 649 946 4436 between 6.00pm and 7.00pm EST (2200 - 2300GMT).
E-mail: info@vp5jm.com *or* jody@tciway.tc

CX6VM guest shack.

three more people.

Equipment available includes Kenwood TS-850SAT and TS-450SAT transceivers, a Drake L7 linear amplifier, a Gonset amplifier, an MFJ 784B DSP filter and a Notebook loaded with amateur radio software.

Outside there are two towers, 118ft and 98ft tall. The antenna system consists of a Hy-Gain TH-7DX 7-element beam for 10, 15 and 20m, a Cushcraft 40-2CD 2-element Yagi for 40m, a Cushcraft A3WS beam for 12, 17 and 30m, Cushcraft XM-510 5-element monobander for 10m. Soon there will also be monobanders for 15 and 20m (ask by e-mail at any time because the station is in continual development), arrays of delta loops and other wire

The 118 and 98ft towers at CX6VM.

UNITED KINGDOM

See England, Jersey, Scotland and Wales.

URUGUAY

CX6VM station, Melo, Cerro Largo Province
Website: www.qsl.net/cx6vm
Jorge Diez, CX6VM, and his family are offering a new rental station just outside the city of Melo, near the border with Brazil. Apart from Jorge's own callsign of CX6VM, the station is well known through its special contest callsigns such as CX5X, CW5W or CW6V.

The shack is 6 x 4 metres in size (58 sq ft) with one bed and also has a kitchenette and bathroom. There is also a guest house with accommodation for

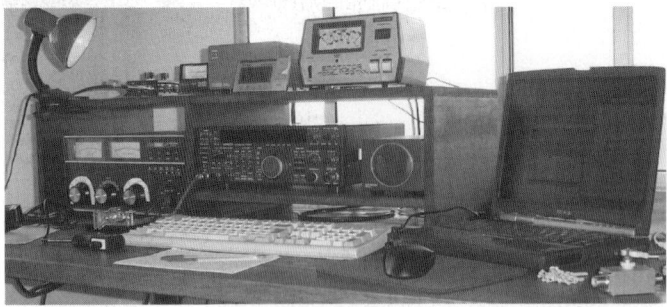

The CX6VM operating position.

SteppIR Yagi at CX2AQ home station.

antennas for the low bands, and long Beverage receive antennas for EU and USA.

A back-up generator is available if there should be a power cut.

The rates vary depending on the number of visitors, whether they use the station equipment or take their own, and whether it is a major contest weekend or not. By way of illustration, for one operator staying for one week outside a major contest period, EUR 700 euros (approx GBP £504) is charged. Please contact Jorge by e-mail for a quote.

Contact: Jorge Diez, CX6VM, Remigio Castellanos 474, Melo, CP 37000 - Cerro Largo, Uruguay.

E-mail: cx6vm.jorge@adinet.com.uy

CX2AQ 'home station', Montevideo
CX2AQ 'contest station'
CX2AQ 'weekend QTH'
Website: www.qsl.net/cx2aq
If you are visiting Montevideo and wish to operate radio, Richard 'Ron' Serván,

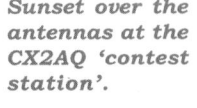

Sunset over the antennas at the CX2AQ 'contest station'.

CX2AQ, offers the use of his station. No accommodation is available, but Ron recommends that you stay at either the Sheraton or the Hotel Cala Di Volpe, both of which are close to his home. Ron will also help you to take out a CX licence. The CX2AQ 'home' station includes an Icom IC-756ProII and Heathkit SB-220 linear amplifier to a SteppIR 3-element beam for 20, 17, 15, 12, 10 and 6m.

Ron also has a contest station 40km north of Montevideo with accommodation for four people. It is equipped with a Yaesu FT-1000MkV, full legal power amplifier and five towers with 10, 15 and 20m monoband Yagis, a tribander and wires.

The CX2AQ home station.

Ron, CX2AQ, at his 'weekend QTH'.

Finally, Ron can offer his 'weekend QTH', located 150km north-east of Montevideo with accommodation for two people. This is a low-band station with an Icom IC-706MKII and antennas for 30, 40 and 80m.
Contact: Richard 'Ron' Serván, CX2AQ, Parva Domus 2521, Montevideo 11300, Uruguay.
E-mail: cx2aq@internet.com.uy

USA

See Alaska, Hawaii, US Virgin Islands.

US VIRGIN ISLANDS

Radio Reef, KP2M station, St Croix (IOTA NA-106)
Website: www.radioreef.com
Radio Reef is located on the north side of the island of St Croix, approximately four miles west of Christiansted. The accommodation consists of 950 sq ft. of living area with views of the Caribbean and a verandah for relaxing, having a barbecue or just enjoying the views of St Thomas, St John and the British Virgin Islands.

The air-conditioned master bedroom features a king bed and separate access to the veranda, while the living room and entertainment area features a TV and comfortable furnishings. The sofa has a queen sleeper should an extra bed be needed. The kitchen is fully furnished with new appliances. High-speed wireless Internet is also available.

The 'Radio Reef DXers' club station has the callsign KP2M and guests are free to use that call or the KP2 prefix with their own callsign. The station consists of a Yaesu FT-1000MP trans-

View from Radio Reef over the Caribbean.

What a take-off! View over the Caribbean from the top of the tower at Radio Reef.

Force 12 Magnum 620/340 for 20 and 40m, and KLM 6m beam.

Radio shack at Tynrhos.

The comfortable shack at Radio Reef.

ceiver and Alpha 89 linear amplifier. Antennas include a Force 12 Magnum 620/340 (6 elements on 20m, 3 elements on 40m) and KLM 6m beam on a crank-up tower as well as a Force12 4BA (12 elements on 10, 12, 15, 17m) beam on a free-standing Rohn 45 tower.

Weekly rates (based on Wednesday to Wednesday occupancy for one or two persons): October to May, contest weekend $1450, non-contest weekend $1150; June to September: negotiable. **Contact:** Brad (K9BZ/KP2) and Donna Zuehlke, Box 946, Christiansted, USVI 00821, USA.
E-mail: info@radioreef.com

simple and economical accommodation and which caters especially for divers. Owned and operated by Chris, GW4VAG, and Val Green, radio amateurs staying there are welcome to use the antennas and, under supervision, the radio equipment.

Tynrhos is located about six miles (10km) west of Pwllheli in the county of Gwynedd and two miles (3km) from the villages of Abersoch and Llanbedrog. St Tudwal's Islands (which count as EU-106 for IOTA) are located just off Abersoch, while Bardsey Island (IOTA EU-124) is 10 miles (16km) to the west of Tynrhos.

You can camp in a tent or bring your own caravan. Alternatively, there

WALES

The Tynrhos grounds with wire antennas.

Tynrhos, near Pwllheli, North Wales (IOTA EU-005)
Tynrhos is a camp site with bed and breakfast establishment, providing

The caravan with Cushcraft A3WS and A4S beams.

is a seven berth static caravan available to rent, and two 'bunk houses' sleeping a total of 10. There is a covered barbecue area for the use of visitors and full cooked British breakfasts are available.

The antennas at Tynrhos are a Cushcraft A4S four-element beam for 10, 15 and 20m, and above that a Cushcraft A3WS three-element beam for 12 and 17m. There is a dipole for 80m and for 160m a 150m-long horizontal wire loop. A Cushcraft R7 vertical antenna is also available. For 2m, there is a six-element quad and a vertical. Covering most bands, the equipment includes Kenwood TS-870, TS-440 and TM-241E transceivers plus a Kenwood RZ1 scanner.

People renting the caravan also have their own 2m antenna and a PSU. No charge is made for the use of the antennas and radio gear but donations for the Lifeboat Fund and the cost of electricity are requested.

The caravan has its own toilet and shower and is fully equipped with pillows, sheets, cutlery, pots and pans, TV, radio, microwave cooker, electric cooker and fridge. Hire charges are £175 a week except in June, July and August when it is £185 a week. The caravan can also be hired at £45 per night for short stays.

The bunkhouses are like Portacabins. One has two separate rooms, with three bunks and two bunks, while the other bunkhouse has two sets of double bunks and a 'put-you-up' bed. Both have cooking facilities with pots and pans, cutlery and radio and TV. Bunkhouse rental rates are £7.50 a night except in June, July and August when they are £8 a night.

In addition there is plenty of camping space with room for car and boat parking. The site has flush toilets, hot showers and fridge and freezer for campers' use. Camping costs £4.50 per person per night with discounts for longer stays.

Full British breakfasts are available at £5.75 each.

Contact: Chris (GW4VAG) and Val Green, Tynrhos, Nanhoron, Pwllheli, Gwynedd LL53 7PS, Wales, UK.
Tel: +44 1758 740712.
E-mail: tynrhosdiving@btinternet.com

Cefn Gribyn, Anglesey (Ynys Môn) (IOTA EU-005)
Websites: www.anglesey-holiday.biz
Cefn Gribyn is a traditional Welsh farmhouse with holiday cottage standing in grounds of around 3.5 acres. It offers high-quality fully-furnished self-catering accommodation for up to four people. Cefn Gribyn is set in rural countryside near the village of Carmel, which in turn is near Llannerchymedd on the island of Anglesey.

The cottage (a non-smoking environment) has two bedrooms (one twin, one double), a large living/dining room, kitchen with electric stove, fridge-freezer and all utensils, bathroom with bath and shower, central heating and

From left to right: Cushcraft R7 vertical, 2m 6-element quad and HF tower with A3WS and A4S beams at Tynrhos.

Cefn Gribyn farmhouse.

Cefn Gribyn holiday cottage.

Tower and beams at Cefn Gribyn.

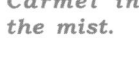

Carmel in the mist.

large gardens.

The shack is in a converted barn separate from the house and includes an Alinco DX-77 transceiver to a home-brew GS35b linear amplifier, a Yaesu FT-101ZD, and a Yaesu FT-2800M 60W 2m FM transceiver. The antennas are a Telrex 3-element monoband Yagi for 20m and a 5-element Yagi for 2m on a 45ft tower, with wire dipoles for 160, 80 and 40m. Broadband Internet is available to either Microsoft or Unix systems. The shack also features a table for electronics work with some modest test gear and a reasonable stock of small components. There is also an outbuilding with tools for mechanical work.

There is an enormous amount to see and do in the area. The cottage is about nine miles from Holyhead on Holy Island (IOTA EU-124) with South Stack Lighthouse, colonies of puffins and, for plane spotters, RAF Valley. The Holyhead ferry will take you to Ireland in about 2.5 hours. Anglesey itself has miles of deserted beaches and the town of Beaumaris with its moated castle and views of Puffin Island. A 25-minute drive crossing the Menai Suspension Bridge takes you to the spectacular Snowdonia national park and the historic town of Caernarfon with its 14th century castle.

Rental of Cefn Gribyn is GBP £250 per week (plus a £50 security bond which is refundable at the end of your stay).

Contact: Les Hayward, MW0SEC, Cefn Gribyn,, Carmel, Llannerchymedd, Isle of Anglesey LL71 7BU, North Wales, United Kingdom.
Tel: +44 1248 470606.
E-mail: les@corfe-castle.demon.co.uk

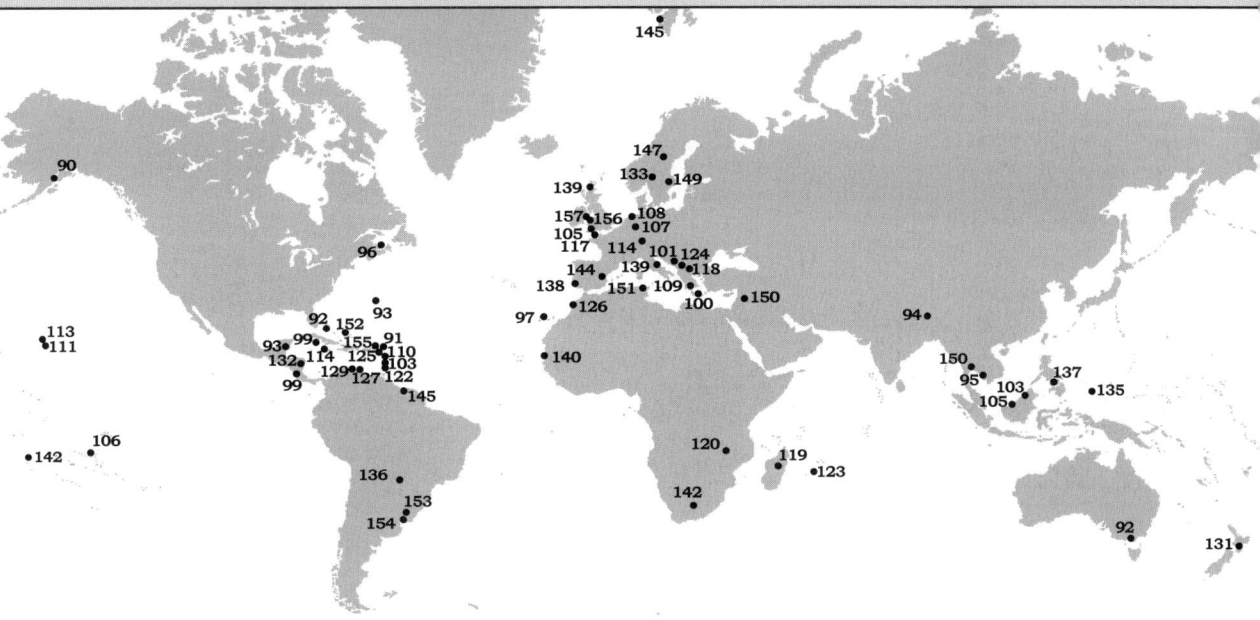

Page numbers of Rental Stations featured in the book.

Index